Band 21 · Mai 1985 **5**

isotopenpraxis

AKADEMIE-VERLAG

ISSN 0021-1915 Isotopenpraxis, Berlin 21 (1985) 5, 153—192

EVP 9,— M

An unsere Autoren

Die Zeitschrift veröffentlicht Beiträge zu folgenden Themen: Kontrolle und Analyse radioaktiver und stabiler Isotope, Anwendung radioaktiver und stabiler Isotope in allen Gebieten der Forschung und Technik, der Industrie, Landwirtschaft und Medizin, Herstellung radioaktiver und stabiler Isotopen sowie von markierten Verbindungen, kernelektronische Geräte für die Isotopenanwendung (Dickenmesser, Dichtemesser u. ä.) Detektoren für isotopentechnische Arbeiten und den Strahlenschutz in Isotopenlaboratorien, Meßtechnik und Strahlenschutz, Dosimetrie in Isotopenlaboratorien, kerntechnische Geräte für die Isotopenanwendung (Beton, Bestrahlungsanlagen u. ä.), Strahlenchemie, Radiochemie, ökonomische und perspektivische Fragen, Fragen der Weiterbildung.

1. Aufnahme von Manuskripten

Zur Veröffentlichung angenommen werden nur druckreife, noch nicht anderweitig publizierte Beiträge in deutscher, englischer oder russischer Sprache zum Themenbereich der Zeitschrift und zwar

Originalarbeiten
Kurze Mitteilungen
Aus der Praxis
Zusammenfassende Berichte
Tagungsberichte
Informationen.

Sind der Beitrag oder Teile daraus bereits veröffentlicht oder ist eine anderweitige Veröffentlichung vorgesehen, so muß dies der Redaktion mit genauer Angabe der Zweitveröffentlichung mitgeteilt werden. Mit der Annahme des Manuskriptes behält sich der Verlag das Recht der Vervielfältigung, Verbreitung und Übersetzung innerhalb der gesetzlichen Schutzfrist vor. Auszugsweiser Nachdruck oder photomechanische u. a. Wiedergabe sind an eine schriftliche Genehmigung des Verlages gebunden. Manuskripte, Zeichnungen und Photos von veröffentlichten Arbeiten werden nur auf den besonderen und bis zum Zeitpunkt der Veröffentlichung mitgeteilten Wunsch des Autors zurückgesandt.

Bei der unverlangten Einreichung von Manuskripten wird empfohlen, eine Befürwortung eines der Herausgeber oder Redaktionsausschußmitglieder beizufügen.

Manuskripte sind einzureichen bei der Redaktion, 1086 Berlin Leipziger Str. 3—4.

2. Form der Manuskripte

Die Autoren werden gebeten:

2 Manuskripte, eines davon in Original-Maschinenschrift, einseitig und mit 2zeiligem Abstand einzureichen; gut leserliche Photokopien sind zulässig; auf dem 1. Blatt (unter Weglassung der Bezeichnung)

den Titel der Arbeit
Vorname-Initialen und Namen sämtlicher Autoren
gegebenenfalls die Institution, in der sie tätig sind
eine kurze Zusammenfassung der Arbeit von höchstens 15 Zeilen
Keywords in alphabetischer Reihenfolge und
als Fußnote die Anschrift der Institution anzugeben;

wenn möglich, Zusammenfassung, Titel und Autoren *englisch* (bei deutschem oder russischem Manuskript) bzw. *deutsch* (bei englischem Manuskript) 2fach beizufügen;

Literaturhinweise (fortlaufend numeriert), Fußnoten, Tabellen, Abbildungen (in 2—3facher Veröffentlichungsgröße), Abbildungsunterschriften (getrennt von den Abbildungen mit Zeichenerklärungen, Quellenangaben, falls erforderlich, und gegebenenfalls mit einem Vermerk für besondere Wünsche bezüglich der Abbildungsgröße), private und Dienstanschrift der Verfasser mit akademischen Titeln und ausgeschriebenen Vornamen auf besonderen Blättern ans Ende des Manuskriptes zu stellen;

Wünsche für den Kolumnentitel (Seitenüberschrift) bekanntzugeben;
Absätze durch Einzüge von mindestens 3 Anschlägen kenntlich zu machen;
die Maßeinheiten des SI-Systems („Systeme International") zu verwenden;
Korrekturen im Manuskript unmittelbar an der zu korrigierenden Stelle anzubringen;

Formeln deutlich lesbar und möglichst groß einzutragen;
Farbabbildungen und nachträgliche Änderungen oder Ergänzungen in Abbildungen zu vermeiden;

hervorzuhebende Textstellen zu unterstreichen, und zwar halbfetten Satz mit einer geraden schwarzen Linie, kursiven Satz (Schrägschrift) mit einer blauen Wellenlinie (alle Eigennamen). Sperrungen gestrichelt;
besondere Buchstaben durch farbige Unterstreichung zu kennzeichnen, und zwar Formelzeichen in

Griechisch-Kursiv ($\alpha, \beta, \ldots$) rot,
Kursiv ($A, B, \ldots$) blau
halbfetter Kursiv ($\boldsymbol{A}, \boldsymbol{B}, \ldots$) 2 Linien blau (Vektoren),
halbfetter Grotesk ($\mathbf{A}, \mathbf{B}, \ldots$) 2 Linien violett (Tensoren);

bei Literaturangaben auf Vollständigkeit zu achten. In nachstehender Reihenfolge müssen angegeben sein:

a) bei *Zeitschriften*: Vorname-Initialen und Namen sämtlicher Autoren, Titel der Zeitschrift (nach „Periodica Chimica" oder möglichst wenig gekürzt), Bandnummer (unterstrichen), Jahreszahl (in Klammern), Heftnummer (nur bei Zeitschriften, die jedes Heft mit S.1 beginnen) sowie die Seitenzahl, auf der die zitierte Arbeit beginnt.

b) bei *Büchern*: Initialen der Vornamen und Name des Autors (Herausgeber werden durch den Zusatz „ed." gekennzeichnet), voll ausgeschriebener Titel des Buches in der Originalsprache, evtl. Bandnummer, voll ausgeschriebener Name des Verlages, Nummer der Auflage (außer der ersten), Erscheinungsort(e), Erscheinungsjahr (hier nicht in Klammern setzen!) und evtl. Kapitelnummer, Seitenzahl u. ä.

c) bei *Sammelbänden, Tagungsberichten*: Es soll zusätzlich der Titel des zitierten Beitrags angegeben werden. Die Seitenzahl muß unbedingt angegeben werden. Aus dem Zitat muß hervorgehen, ob es sich um einen Tagungssammelband oder um einen noch nicht veröffentlichten Vortrag handelt. Zwischen Tagungsort und Erscheinungsort des entsprechenden Tagungsberichts ist zu unterscheiden.

3. Korrekturen

Der Autor erhält zwei Umbruchabzüge, von denen einer möglichst schnell korrigiert an die Redaktion zurückgereicht werden soll. Auf Wunsch des Autors kann die Korrektur auch von der Redaktion gelesen werden.

Durch nachträgliche Manuskriptänderungen verursachte Korrekturkosten gehen, sofern sie 10% der Satzkosten übersteigen, zu Lasten des Autors.

Der Autor erhält 50 Sonderdrucke kostenlos. Weitere Sonderdrucke können bis zur Rückgabe der Korrekturabzüge auf Kosten des Autors bei der Redaktion bestellt werden.

Bezugsmöglichkeiten

Bestellungen sind zu richten

- **in der DDR** an den Postzeitungsvertrieb unter Angabe der Kundennummer des Bestellenden oder an den AKADEMIE-VERLAG Berlin, DDR-1086 Berlin, Leipziger Straße 3—4, PF-Nr. 1233;
- **im sozialistischen Ausland** an eine Buchhandlung für fremdsprachige Literatur oder an den zuständigen Postzeitungsvertrieb;
- **in der BRD und Berlin(West)** an eine Buchhandlung oder an die Auslieferungsstelle KUNST UND WISSEN, Erich Bieber OHG, Wilhelmstraße 4—6, D-7000 Stuttgart 1;
- **in den übrigen westeuropäischen Ländern** an eine Buchhandlung oder an die Auslieferungsstelle KUNST UND WISSEN, Erich Bieber GmbH, Dufourstraße 51, CH-8008 Zürich;
- **im übrigen Ausland** an den Internationalen Buch- und Zeitschriftenhandel; den Buchexport Volkseigener Außenhandelsbetrieb der Deutschen Demokratischen Republik, DDR-7010 Leipzig, Postfach 160; oder an den AKADEMIE-VERLAG Berlin, DDR-1086 Berlin, Leipziger Straße 3—4.

Zeitschrift „Isotopenpraxis"

Herausgeber: Zentralinstitut für Isotopen- und Strahlenforschung der Akademie der Wissenschaften der DDR.
Verlag: Akademie-Verlag Berlin, DDR-1086 Berlin, Leipziger Straße 3—4, PF-Nr. 1233; Fernruf: 2236221 oder 2236229; Telex-Nr.: 114420;
Bank: Staatsbank der DDR, Berlin, Kto.-Nr.: 6836-26-20712.
Chefredakteur: Prof. Dr. H. Hübner, Leipzig.
Anschrift der Redaktion der Zeitschrift Isotopenpraxis: Prof. Hübner, DDR-7010 Leipzig, Karl-Marx-Platz, Hauptpostamt, PSF 370; Fernruf: 281018.
Veröffentlicht unter der Lizenznummer 1527 des Presseamtes beim Vorsitzenden des Ministerrates der Deutschen Demokratischen Republik.
Gesamtherstellung: VEB Druckerei „Thomas Müntzer", DDR-5820 Bad Langensalza.
Erscheinungsweise: Die Zeitschrift „Isotopenpraxis" erscheint monatlich. Die 12 Hefte eines Jahrganges bilden einen Band. Bezugspreis je Band 240,— DM zuzüglich Versandspesen; Preis je Heft 20,— DM. Der gültige Jahresbezugspreis für die DDR ist der Postzeitungsliste zu entnehmen.
Bestellnummer dieses Heftes: 1079/21/5.

AN (EDV) 12831
00900

Isotopenpraxis 21 (1985) 5, S. 153 bis 159

Some Geochemical Applications of a Semi-Empirical Approach for Estimating Distribution Coefficients of a Series of Cations between Liquid Aluminosilicates and their Crystallization Products

K. Wetzel

(Central Institute of Isotope and Radiation Research of the Academy of Sciences of G.D.R.)[1])

The distribution coefficients of alkali, alkaline earth, rare earth and a series of other elements between solid and liquid natural aluminosilicate systems can be estimated by a semi-empirical equation which can be derived by extension of Schuetze's incremental method for calculating oxygen isotope effects in silicate systems. This equation is used for investigating the partial melting of mantle wedge and subducted oceanic crust as potential sources of magmas giving rise to continental growth and for evaluating the history of igneous rocks. A series of geochemical criteria are derived for attributing igneous rocks to common parent magmas and for contributing to the estimation of the length of intrusive pathways. Particular attention is devoted to the evaluation of the geochemical behaviour of the rare earth elements.

Die Verteilungskoeffizienten der Alkalimetalle, der Erdalkalimetalle, der Seltenerdmetalle und einer Reihe weiterer chemischer Elemente zwischen festen und flüssigen natürlichen Aluminosilikatsystemen können mit Hilfe einer halbempirischen Beziehung berechnet werden, die durch Erweiterung der Inkrementenmethode von Schütze zur Berechnung der Isotopieeffekte des Sauerstoffs in Silikatsystemen entsteht. Diese Beziehung wird zur Untersuchung des partiellen Schmelzens von Material des Oberen Mantels bzw. der in den Mantel abtauchenden ozeanischen Kruste als potentielle Quellen der das kontinentale Wachstum bewirkenden Magmen und zur Beurteilung der Entstehungsweise magmatischer Gesteine herangezogen. Es wird eine Reihe geochemischer Kriterien abgeleitet zur Klärung der Frage der Zugehörigkeit von Intrusivkörpern zu gemeinsamen Stammagmen und zur Beurteilung der Länge des Intrusionsweges von Magmen. Besonders eingegangen wird auf das geochemische Verhalten der Seltenerdelemente.

Keywords

alkaline earth metals; crystal structure; crystallization; distribution functions; geochemistry; geology; granites; igneous rocks; isotope ratio; models; oxygen 18; rare earths; separation processes; thermodynamics

1. Introduction

(Si, Al)O_4 tetrahedrons are the fundamental constituents of all aluminosilicates. The fact that the bond strengths between oxygen on one hand and silicon or tetrahedrally coordinated aluminium on the other hand dominate over the bond strengths between oxygen and all other cations gives rise to an extensive homogenity and continuity of structure and chemical composition of aluminosilicate rocks. This suggests the extension of *Schuetze*'s ^{18}O indices for calculating oxygen isotope equilibria between silicatic minerals and rocks [1 – 3] also on the calculation of the distribution coefficients

$$D = \frac{\text{concentration in the solid phase}}{\text{concentration in the liquid phase}}$$

of chemical elements between magmas and their solidification products:

$$\lg D = k + S \cdot \Delta I \qquad (1)$$

where k and S are coefficients to be determined empirically and $\Delta I = I_1 - I_s$ is the difference of the ^{18}O indices of the magma (I_1) and its crystallization products (I_s). The ^{18}O indices can be calculated by the equation

$$I = \frac{1}{y}(0{\cdot}977x + 0{\cdot}645z + 0{\cdot}0155w + 4{\cdot}092) \qquad (2)$$

where x, y and z are the formula indices of Si, O and not tetrahedrally coordinated Al in a chemical formula $M^{n}_{w/n}Al_z[Si_xAl_{4-x}O_y]$ of the silicate. The mean electrical charge of cations other than Si and Al is denoted by n, their formula index by $w = 2y - x - 3z - 12$ [4]. The ^{18}O index is a measure of the geometric configuration and the bond relationships of that oxygen ions the lattice of which hosts the cations not in tetrahedral positions. (Contrary to isotope fractionation coefficients distribution coefficients of chemical elements will not become equal to unity if there is no difference in the geometric configuration and the bond relationships of the O^{2-} lattice: If $\Delta I = 0$, $\lg D$ in equation (1) is equal to k and not equal to zero like in *Schuetze*'s formula for oxygen isotope equilibria.)

The coefficients k and S already calculated for the alkali, the alkaline earth, the transition elements of the first period and the rare earth elements [2, 3] are summarized in Tab. 1. We emphasize that the argument of homogenity of structure and chemical composition of aluminosilicates only holds for rocks, not for single minerals. Therefore equation (1) is only valid for whole rock systems, not for single minerals or for mineral matrix systems. It should also be repeated that k and S are more or less dependent on the particular crystallization conditions (temperature, pressure, rate of solidification and – as we suppose – especially water content of the magma). The values for k and S in Tab. 1 are valid for average crystallization conditions of continental magmas.

[1]) Address: DDR-7050 Leipzig, Permoserstraße 15

Tab. 1. Coefficients k and S in equation (1), regression coefficient r, mean percent deviation $\bar{\Delta}$ of calculated distribution coefficients from those derived from *Clarke* values in igneous rocks and concentrations (in ppm apart from elements given as oxides which are in weight percent) in oceanic crust, upper mantle and continental crust. c'_{KK} is the concentration in the continental crust, if partial melting of mantle wedge is predominant, c''_{KK} the concentration in the continental crust mainly growing due to partial melting of downgoing slab.

element	k	S	r	$\bar{\Delta}$, %	concentration in				
					oceanic crust	upper mantle	continental crust	c'_{KK}	c''_{KK}
					ppm	ppm	ppm	ppm	ppm
Li	−0·071	−1·79	−0·55	16	10	0·5 (1·4)	20	1·2 (3·3)	16
Na_2O	+0·102	−2·54	−0·94	5·3	2·23	0·28	2·5	0·58	3·01
K_2O	−0·239	−3·14	−0·68	27	0·82	0·04	2·1	0·26	2·14
Rb	−0·189	−3·09	−0·87	13	32	2 (4·2 or 0·7)	90	10·1 (21·3 or 3·5)	75
Cs	+0·0099	−5·17	−0·96	10	1·0	0·1	3	0·71	2·4
Be	+0·321	−11·4	−0·98	13	0·5	0·2	2·8	7·6	1·7
MgO	+0·235	+0·983	+0·57	7·9	6·34	39·0	4·9	15·5	3·28
CaO	+0·229	+0·490	+0·83	2·3	10·42	3·25	7·2	1·59	6·06
Sr	+0·202	−4·70	−0·78	24	440	19	375	72·6	168
Ba	+0·056	−4·59	−0·77	22	300	1 (38)	425	1·6 (60·4)	252
Sc	+2·14	−6·71	−0·90	3·8	38	12·1	22	1·2	1·6
TiO_2	+0·047	+1·95	+0·90	5·6	2·03	0·09	0·7	0·038	1·27
V	+0·007	+3·04	+0·92	8·3	275	100	135	31	148
Cr	−1·77	+26·5	—	—	200	2800	100	6·3	28·6
MnO	+0·041	+1·82	+0·79	8·7	0·176	0·11	0·10	0·050	0·114
FeO	+0·112	+1·16	+0·62	9·3	11·59	8·0	7·76	3·96	7·45
Co	−0·95	+18·9	—	—	45	200	25	1·28	5·91
Ni	−0·445	+11·5	—	—	160	2000	75	67·3	35·7
Cu	−0·109	+5·57	+0·99	4·3	100	20	55	3·0	40
Zn	−0·023	+2·14	+0·92	6·0	130	30 (56)	70	13·9 (25·9)	90
Y	−0·085	+0·366	+0·81	2·9	34	2·7	33	2·85	37·7
La	+0·178	−7·89	−0·77	34	4·84	1·74	30	23·9	13·2
Ce	+0·108	−5·69	−0·79	26	15·5	4·85	60	33·6	34·6
Pr	+0·090	−5·14	−0·85	18	2·53	0·71	8·2	4·16	5·35
Nd	+0·055	−4·04	−0·89	13	13·2	2·53	28	10·5	24·8
Sm	+0·020	−2·93	−0·96	4·5	4·25	0·55	6	1·62	7·06
Eu	+0·253	−3·23	−0·89	9·9	1·63	0·24	1·2	0·46	1·90
Gd	−0·0149	−1·83	−0·95	1·5	6·10	0·72	5·4	1·50	8·90
Tb	−0·050	−0·734	+0·83	1·3	1·13	0·145	0·9	0·214	1·44
Dy	−0·067	−0·184	+0·68	3·4	3 ?	0·77	3	0·96	3·57 ?
Ho	−0·085	+0·366	+0·89	3·4	1·37	0·125	1·2	0·132	1·52
Er	−0·102	+0·916	+0·87	2·9	3·86	0·34	2·8	0·302	3·98
Tm	−0·120	+1·466	+0·90	4·0	0·61	0·047	0·48	0·035	0·585
Yb	−0·137	+2·016	+0·63	12	3·25	0·39	3	0·246	2·89
Lu	−0·155	+2·567	+0·71	11	0·52	0·039	0·5	0·021	0·43

2. Application for investigating the source of continental growth

As we now can approximately calculate the distribution coefficients of 35 chemical elements between solid and liquid natural aluminosilicats of any chemical composition we will try to contribute to answering a fundamental question of plate tectonics: Is partial melting of upper mantle between convergent continental margins and subducted oceanic crust or is partial melting of subducted oceanic crust itself the source of magmas giving rise to continental growth? The two models of the geochemical mechanism of continental growth shall be used to calculate the concentrations of the 35 elements in magmas having been generated in one or the other way and to compare these concentrations with the average concentrations c_{KK} of these elements in the continental crust.

If partial melting of mantle wedge is the main process, the following equation holds for the concentration c'_{KK} in the continental crust:

$$c'_{KK} = \frac{c_{OM}}{10^{(k+0\cdot1668\cdot S)}} \tag{3}$$

where c_{OM} is the concentration in the upper mantle and k and S are the coefficients in equation (1). If continental growth is predominantly due to partial melting of subducted oceanic crust, the concentration c''_{KK} in the continental crust is ruled by the equation:

$$c''_{KK} = \frac{c_{OZK}}{0\cdot172 + 0\cdot815\cdot10^{(k+0\cdot1016\cdot S)}} \tag{4}$$

where c_{OZK} is the concentration in the oceanic crust [5].

The last two columns of Tab. 1 show c'_{KK} and c''_{KK} calculated by the equations (3) and (4). As concentrations c_{OZK} data for tholeiitic or oceanic basalts given by *Wedepohl* [6] have been used. With respect to c_{OM} we follow *Hutchison*'s Lherzolite model [7] as far as data are given. The other values for c_{OM} are taken from data for upper mantle or peridotites or chondrites, respectively, given by *Beus* [8] and *Wedepohl* [6]. (If there are severe discrepancies between such data we take two or three values into account (figures in brackets) in favour of the partial melting of mantle wedge.) As for the concentrations c_{KK} in the continental crust we follow the data of *Taylor* [9] as far as possible, the concentrations of main elements being taken from *Ronov* and *Jaroshevskij*'s [10] estimate.

As Tab. 1 shows, there is by far a better agreement between c''_{KK} and c_{KK} than between c'_{KK} and c_{KK}. Only in the case of 2 of 36 elements, namely in the case of nickel and lanthanum, partial melting of mantle wedge gives a better explanation of the *Clarke* values of the continental crust than partial melting of downgoing slab. In addition to that the agreement between c''_{KK} and c_{KK} is within the uncertainties of *Clarke* values in the case of about 20 elements, while the discrepancies between c'_{KK} and c_{KK} are severe in more than 20 cases.

Thus we can conclude that partial melting of subducted oceanic crust as the main source of magmas effecting continental growth is supported not only be the geochemistry of alkali, alkaline earth and transition elements of the first period of the periodic table, as we had already shown [5], but also by the geochemistry of the rare earth elements.

3. Applications for investigating the history of igneous rocks

3.1. *Behaviour of element pairs during magmatic processes*

Equation (1) also allows to describe the behaviour of element pairs M_1/M_2 during fractional crystallization:

Subtraction of the single equations

$$\lg D_1 = k_1 + S_1 \cdot \Delta I$$

and

$$\lg D_2 = k_2 + S_2 \cdot \Delta I$$

yields:

$$\lg D_1 - \lg D_2 = (k_1 - k_2) + (S_1 - S_2)\,\Delta I \qquad (5\,\text{a})$$

or

$$\lg D_1/D_2 = k_{1/2} + S_{1/2} \cdot \Delta I \qquad (5\,\text{b})$$

with $k_{1/2} = k_1 - k_2$ and $S_{1/2} = S_1 - S_2$

That means equation (1) also holds for element pairs if the distribution coefficient ratio is used instead of distribution coefficient and if the differences of the coefficients k and S are applied in the place of the coefficients themselves.

Let us consider the behaviour of element pairs during fractional crystallization. If every portion of the solid phase is and stays in equilibrium with the homogeneous melt from which it crystallized (first type crystallization differentiation), *Rayleigh*'s equation holds:

$$\lg c_1 = \lg c_1^0 + (D_1 - 1) \lg M/M_0$$

$$\lg c_2 = \lg c_2^0 + (D_2 - 1) \lg M/M_0\,,$$

where c_1^0 and c_2^0 are the concentrations of the two elements 1 and 2 at the very beginning of crystallization, M_0 the amount of the melt in this moment and c_1 and c_2 the concentrations in the residual melt of the amount M. Subtraction of these equations yields

$$M/M_0 = 10^{\frac{1}{D_2 - D_1} \lg \frac{c_1^0/c_2^0}{c_1/c_2}}$$

which can approximately be substituted by

$$M/M_0 \approx 10^{\frac{1}{1 - D_1/D_2} \lg \frac{c_1^0/c_2^0}{c_1/c_2}}$$

if $D_2 \approx 1$. Comparison of this equation with *Rayleigh*'s equation

$$M/M_0 = 10^{\frac{1}{1 - D} \lg \frac{c^0}{c}}$$

leads to the conclusion that element pairs approximately behave like elements during fractional crystallization if concentration is replaced by concentration ratio and if distribution coefficient is replaced by distribution concentration ratio.

It can be shown that the same is true for the case that the entire solid phase is in thermodynamic equilibrium with the residual melt (second type crystallization differentiation). Equilibrium conditions

$$\frac{c_1^s}{c_1} = D_1 \quad \text{and} \quad \frac{c_2^s}{c_2} = D_2$$

have to be combined with the balance equations

$$M/M_0 \cdot c_1 + (1 - M/M_0)\, c_1^s = c_1^0$$

and

$$M/M_0 \cdot c_2 + (1 - M/M_0)\, c_2^s = c_2^0$$

in order to obtain

$$\frac{D_1}{D_2} = \frac{c_1^0 - M/M_0 c_1}{c_2^0 - M/M_0 c_2} : c_1/c_2$$

which becomes identical with

$$\frac{D_1}{D_2} = \frac{c_1^0/c_2^0}{c_1/c_2}$$

if $M/M_0 \ll 1$ and $M/M_0 c \ll c^0$. (c_1^s or c_2^s, respectively, means the concentrations in the solid phase.) Thus, under certain conditions being fulfilled in most cases of practical importance, equation (1) also holds for element pairs if the concentration ratio is used instead of the concentration and if the differences of the coefficients k and S are applied in the place of the coefficients themselves.

Equation (5) is a tool for selecting element pairs with desirable behaviour during fractional differentiation.

Thus it is opportune for evaluating the origin of magmas and the genesis of igneous rocks to get element pairs with $S_1 \approx S_2$ ($S_{1/2} \approx 0$) the D_1/D_2 values of which are approximately independent on the chemical composition of the magma as equation (5) shows. Elements fulfilling the condition $|S_{1/2}| \leqq 0{\cdot}7$ are assigned in Tab. 2. (If $|S_{1/2}| \leqq 0{\cdot}7$, the D_1/D_2 value of gabbroic rocks ($\Delta I = 0{\cdot}091$) never deviates from the value of granitic rocks ($\Delta I = 0$) by more than 16% ($10^{0\cdot7\cdot0\cdot091} = 1{\cdot}16$!).

The substitution of single elements by element pairs of Tab. 2 using equation (5) is particularly important for granitic magmas, because the ΔI of granitic rocks are small differences of large figures I_l and I_s thus the term $S \cdot \Delta I$ in equation (1) becoming the more inaccurate the larger S is. If $|k_{1/2}| \leqq 0{\cdot}065$, D_1/D_2 becomes smaller than 1·16 or larger than 0·86, i. e. $D_1/D_2 \approx 1$. The concentration ratios of such element pairs (Li/Gd, K/Rb, Cs/Ba, Mg/Ca, Ba/Nd, Ti/Mn, Mn/Zn) are approximately not

Tab. 2. Element pairs with $|S_{1/2}| \leqq 0{\cdot}7$, their $k_{1/2}$ and $S_{1/2}$ values, their distribution coefficient ratio $(D_1/D_2)_{\Delta I=0}$ for granites, the product of their regression coefficients r_1 and r_2 with respect to equation (1) and the mean percent deviations Δ_1 and Δ_2 of calculated D_1 and D_2 values from those derived from *Clarke* values in igneous rocks

	$k_{1/2}$	$S_{1/2}$	$(D_1/D_2)_{\Delta I=0}$	$r_1 \times r_2$	Δ_1/Δ_2
Li/Gd	−0·056	+0·040	0·879	0·52	16/1·5
Na/K	+0·341	+0·600	2·19	0·64	5·3/27
Na/Rb	+0·291	+0·550	1·95	0·82	5·3/13
Na/Sm	+0·082	+0·390	1·21	0·90	5·3/4·5
K/Rb	−0·050	−0·050	0·89	0·59	27/13
K/Sm	−0·259	−0·210	0·551	0·65	27/4·5
Cs/Sr	−0·192	−0·470	0·643	0·75	10/24
Cs/Ba	−0·046	−0·580	0·899	0·74	10/22
Cs/Pr	−0·080	−0·030	0·832	0·82	10/18
Mg/Ca	+0·006	+0·493	1·014	0·47	7·9/2·3
Mg/Fe	+0·123	−0·177	1·33	0·35	7·9/9·3
Mg/Y	+0·320	+0·617	2·09	0·46	7·9/2·9
Ca/Fe	+0·117	−0·670	1·31	0·51	2·3/9·3
Ca/Y	+0·314	+0·124	2·06	0·67	2·3/2·9
Sr/Ba	+0·146	−0·110	1·40	0·60	24/22
Sr/Nd	+0·147	+0·660	1·40	0·69	24/13
Ba/Nd	−0·001	−0·550	1·002	0·69	22/13
Ti/Mn	+0·006	+0·130	1·021	0·71	5·6/8·7
Ti/Zn	+0·070	−0·190	1·17	0·83	5·6/6·0
Ti/Yb	+0·184	−0·066	0·729	0·57	5·6/12
V/Lu	+0·162	+0·473	1·45	0·65	8·3/11
Mn/Fe	−0·071	+0·660	0·849	0·49	8·7/9·3
Mn/Zn	+0·064	−0·320	1·16	0·73	8·7/6·0
Mn/Yb	+0·178	−0·196	1·51	0·50	8·7/12
Fe/Tm	+0·232	−0·306	1·71	0·56	9·3/4·0
Zn/Yb	+0·114	+0·124	1·30	0·58	6·0/12

Tab. 3. Element pairs with $|k_{1/2}| \geqq 0{\cdot}1$ and $|S_{1/2}| \leqq 0{\cdot}7$ and their *Clarke* values (ppm) in Older (ÄG) and Younger Granites (JG) of the Western Ore Mountains and in gneisses. (Pairs involving rare earth elements have been omitted because of the large differences in the *Clarke* values of granites and gneisses.)

	$k_{1/2}$	$S_{1/2}$	Clarke value		
			ÄG	JG	gneisses
Na/K	+0·341	+0·600	24670/37670	25670/38670	25750/21770
Na/Rb	+0·291	+0·550	24670/250	25670/865	25750/100
Cs/Sr	−0·192	−0·470	−/118	−/17	−/376
Mg/Fe	+0·123	−0·177	3133/10800	830/6770	10072/16000
Ca/Fe	+0·117	−0·670	7100/10800	2767/6770	17160/16000
Sr/Ba	+0·146	−0·110	118/450	17/105	376/600

affected by magmatic processes and therefore are a good tool for evaluating other geochemical events than fractional crystallization or partial melting, respectively. Deviations from the constancy of the concentration ratios of such element pairs within igneous rock systems give strong evidence of other than magmatic processes.

The case of element pairs with large $|k_{1/2}|$ values ($|k_{1/2}| \geqq 0{\cdot}1$) will be treated in connection with the search for common parent magmas.

3.2. *Search for common parent magmas*

The capacity of a magma to generate ore deposits is connected with its size and its relations to larger magmatic reservoirs in deeper layers of the crust [11]. The search for common parent magmas of intrusive granitic rocks is therefore not only a theoretical, but also a practical question. The knowledge of distribution coefficients D of single elements between magmas and their solidification products can help to identify such common parent magmas [12] in the following way: The parent magma is treated as a closed system underlying fractional crystallization. Let M_0 be the initial amount of this magma and c_0 the initial concentrations of any minor elements in it. After partial solidification part of the residual molten phase with the amount M is intruded into the overlying rocks. We assume that the amount leaving the parent magma is neglegible in comparison with the remaining amount M. Measurement of element concentrations c in every intrusive rock allows the calculation of M/M_0:

$$M/M_0 = 10^{\left[\frac{1}{1-D} \lg \frac{c_0}{c}\right]} \tag{6}$$

If different chemical elements lead to plausible and consistent M/M_0 values, the existence of a common parent magma may be assumed. An investigation on the Mg and

Ca concentrations in intrusive granites of the Western Ore Mountains gave support to the hypothesis that these granites can be derived from such a common parent magma [12]. We will now try to extend this investigation by involving element pairs in Tab. 2 being distinguished by D_1/D_2 strongly deviating from unity ($|k_{1/2}| \geqq 0{\cdot}1$ or $0{\cdot}79 \geqq D_1 - D_2 \geqq 1{\cdot}26$). Tab. 3 assigns these element pairs together with their *Clarke* values in granites and in gneisses which are supposed to have contamined the intrusive granitic magmas in Western Ore Mountains [13].

Except for Na/Rb all the concentration ratios in Tab. 3 can be severly falsified by contamination with adjacent gneisses. Thus application of equation (6) has to be restricted to Na/Rb.

The results are shown in Tab. 4 in comparison with those obtained using the concentrations of Mg and Ca in the same granites. (ZG 1,2 denotes the so called Intermediate Granites of the Ore Mountains.)

Tab. 4. M/M_0 values of the varistic postkinematic granites of the Western Ore Mountains derived from Mg and Ca concentrations and from Na/Rb concentration ratios

granite	*Clarke* value			M/M_0 from		
	Na	Rb	Na/Rb	Na/Rb	Mg	Ca
ÄG 1	24670	260	94·9	0·79	0·52	0·75
ÄG 2		324	76·1			
ÄG 3		384	64·2	0·84	0·56	0·69
ZG 1, 2	25170	500	50·3	0·77	0·42	0·43
JG 1	25670	838	30·6	0·59	0·84	0·73
JG 2, 3		1166	21·2	0·68	0·66	0·77

The good agreement of the M/M_0 values with one another gives further evidence of the existence of a common parent magma that the granites have been generated from.

It is to some extend an inversion of this idea, if we propose to use the concentrations of those elements or the concentration ratios of those element pairs being distinguished by $D \approx 1$, particularly by $k \approx S \approx 0$, or by $D_1/D_2 \approx 1$, particularly by $k_{1/2} \approx S_{1/2} \approx 0$, respectively, for tracing common parent magmas: Igneous rocks which have been generated from one and the same parent magma should be characterized by an approximate coincidence of the concentrations or concentration ratio of such elements or element pairs provided the concentrations are not affected by contamination with adjacent rocks or by postmagmatic processes. Tab. 5 assigns the elements and elements pairs for which $|k|$ or $|k_{1/2}| \leqq 0{\cdot}065$ and $|S|$ or $|S_{1/2}| \leqq 0{\cdot}7$, i. e. which can be used as criteria for common parent magmas on the proposed way together with the concentrations in granites and in gneisses as typical adjacent rocks as far as possible.

The effects of contamination by adjacent rocks have to be evaluated in dependence on the particular geological situation. In some cases chemical analyses of xenolithes will give good evidence of such effects.

Tab. 6 assigns the results of an application of element pairs of Tab. 5 on the investigation of the varistic postkinematic granites of the Western Ore Mountains, as far as experimental data are available.

The degrees η of contamination agree rather good with each other, but are significantly higher than we have calculated from element concentrations themselves [13].

Tab. 5. Elements and element pairs with $|k|$ or $|k_{1/2}| \leqq 0{\cdot}065$ and $|S|$ or $|S_{1/2}| \leqq 0{\cdot}7$

	k or $k_{1/2}$	S or $S_{1/2}$	concentrations in granites, ppm	concentrations in gneisses, ppm
Y	−0·093	+0·653	34	—
Gd	+0·013	−0·695	9	—
Tb	−0·050	−0·734	2·5	—
Dy	−0·067	−0·184	6·7	—
Li/Gd	−0·056	+0·040	11·3/7·6	−/−
K/Rb	−0·050	−0·50	33400/200	26100/100
Cs/Ba	−0·046	−0·580	5/830	6 ?/600
Mg/Ca	+0·006	+0·493	1600/15800	16700/24000
Ba/Nd	−0·001	−0·55	830/46	600/24
Ti/Mn	+0·006	+0·13	2300/800	6200/500
Ti/Zn	−0·070	−0·19	2300/60	6200/80
Mn/Zn	+0·064	−0·320	800/60	500/80

Tab. 6. Concentration ratios K/Rb, Mg/Ca and Ti/Zn in ÄG and JG granites (see text of Tab. 3) and in grey gneisses (GG) and the calculated degree η of contamination (in mass %) of ÄG by gneisses

	concentration ratio			η, %
	ÄG	JG	GG	
K/Rb	151	45	218	61
Mg/Ca	0·44	0·30	0·59	48
Ti/Zn	30·6	12·3	52·5	55

Since the effects of fractional crystallization on the concentrations are approximately eliminated in the balance calculations presented here we claim that $\eta \approx 55\%$ is probably the more reliable approximation.

3.3. *Search for long intrusive paths*

The capacity of a magma to generate ore deposits is not only connected with its relations to larger magmatic reservoirs in deeper regions of the crust, but also dependent on the length of the intrusive path. The longer this pathway the higher the degree of differentiation by fractional crystallization and the more pronounciated the ore forming capacity should be. Let us consider what equation (1) or equation (5), respectively, can help in recognizing such long pathways. Rocks formed from such magmas should be distinguished by concentration ratios extremely deviating from global average values just of such element pairs the D_1/D_2 values of which strongly deviate from unity. An analysis of the k and S values in Tab. 1 on the basis of equation (5) shows than this is true

in the case of granites ($\Delta I = 0$) for
Eu, Be, Mg, Ca, Sr/Cr, Co, Ni, K,
in the case of granodiorites ($\Delta I = 0{\cdot}019$) for
Mg, Ca, Sr/Cr, Co,
in the case of diorites ($\Delta I = 0{\cdot}074$) for
Co, Ni, Cu/K, Rb, Cs, Be and
in the case of gabbros ($\Delta I = 0{\cdot}091$) for
Ni, Co, Cu/K, Rb, Cs, Be, Ba.

As the Younger Granites (JG) of the Western Ore Mountains are supposed to have been generated from subvolcanic magmas distinguished by a long intrusive pathway [14],

Tab. 7. Concentration ratios of element pairs with D_1/D_2 values in granites strongly deviating from unity in ÄG and JG type granites (see text of Tab. 3), the corresponding ratios in global average granites and $(D_1/D_2)_{\Delta=0}$

	concentration ratio			$(D_1/D_2)_{\Delta I=0}$
	ÄG	JG	granites	
Be/Cr	1·5	4·2	0·21	123
Mg/Cr	653	231	164	101
Ca/Cr	1480	770	486	98
Sr/Cr	24·5	4·8	7·5	94
Eu/Cr	1·5	0·10	0·044	105
Mg/Co	631	244	4100	15·3
Ca/Co	1449	814	12150	15·1
Sr/Co	24	5·0	188	14·2
Eu/Co	1·5	0·11	1·1	16·0
Be/Ni	0·88	7·5	0·67	5·8
Mg/Ni	392	415	513	4·8
Ca/Ni	888	415	1519	4·8
Sr/Ni	15	8·6	23·5	4·4
Eu/Ni	0·91	0·18	0·14	5·0
Be/K	$1{\cdot}9 \times 10^{-4}$	$3{\cdot}9 \times 10^{-4}$	$1{\cdot}25 \times 10^{-4}$	3·6
Mg/K	0·083	0·021	0·096	3·0
Ca/K	0·188	0·072	0·283	2·9
Sr/K	$3{\cdot}1 \times 10^{-3}$	$0{\cdot}44 \times 10^{-3}$	$4{\cdot}4 \times 10^{-3}$	2·8
Eu/K	$1{\cdot}9 \times 10^{-4}$	$0{\cdot}09 \times 10^{-4}$	$0{\cdot}026 \times 10^{-4}$	3·1

their corresponding concentration ratios are compared with the global average concentration ratios of granites.

Tab. 7 indeed shows Mg/Co, Ca/Co, Sr/Co, Eu/Co, Ca/Ni, Sr/Ni, Mg/K, Ca/K and Sr/K ratios of the Younger Granites far below the global average values. (The comparison with the ÄG type granites is less important because of their strong contamination with adjacent rocks.) Thus element pairs with Eu cannnot be recommended probably because of the generally anomalous geochemical properties of this tri- and divalent cation. The dubious behaviour of pairs with Cr is perhaps due to the fact that its concentration in the grey gneisses of the Ore Mountains exceeds the concentrations in JG and ÄG by a factor of 14 or 10, respectively. In conclusion we for the time being only can recommend Mg, Ca, Sr/Co-, Ca, Sr/Ni- and Mg, Ca, Sr/K-ratios at least by a factor of three below average granite values as indicators of long intrusive pathways provided it can be excluded that the deviations are due to contamination by adjacent rocks or by postmagmatic processes.

4. Conclusions

The extended agreement between results obtained by the application of equations (1) and (5) using coefficients k and S calculated on the basis of global *Clarke* values of igneous rocks with geological experience and the fact that k and S can be understood in terms of thermodynamic properties of molten aluminosilicates and their crystallization products [15, 3] give evidence of the following conclusions:

4.1. Clarke values of alkali, alkaline earth and rare earth elements and transition elements of the first period in igneous rocks are mainly determined by thermodynamic equilibria between liquid and solid aluminosilicates. If distribution coefficients D do not strongly deviate from unity this is approximately true independent on wether the fractional crystallization is interrupted by more or less isochemical processes like physical wheathering or metamorphosis or not [15]. (Such strong deviations of D from unity do occur in the case of K, Rb, Cs, Be, Mg, Ca, Sr, Ba, Ti, V, Cr, Co, Ni, Cu and the light rare earth elements in mafic rocks and in the case of Sr in felsic rocks.)

4.2. There is only a limited influence of the crystallization conditions (temperature, pressure, water content, content of other fluids) on the solid-liquid distribution coefficients of the chemical elements investigated here. The most important factor effecting these distribution coefficients is supposed to be the water content of the magma. Thus equations (1) and (5) are valid not only for global *Clarke* values of igneous rocks but — with certain limitations — also for the element concentrations in particular igneous rocks. The most pronounciated deviations from k and S calculated on the basis of global *Clarke* values are to be expected from magmas having extremely high or extremely low water contents in comparison to the average of a magma of the given chemical composition, from ultramafic rocks the *Clarke* values of which have not been involved into the regression calculation for lack of accurate knowledge about the genetic relations of those rocks to the other igneous rocks used, and from Cr, Co and Ni for lack of reliable global *Clarke* values in granodiorites or quartzdiorites, respectively. In spite of the good agreement of distribution coefficients calculated from equation (1) with those estimated on the basis of *Clarke* values of andesites and rhyolites we suppose equations (1) and (5) to be more reliable for plutonic rocks than for volcanic ones. (The high solidification rate of volcanic magmas should give rise to marked deviations from thermodynamic behaviour.) If the application of equations (1) or (5) yields extremely low concentrations in comparison with global *Clarke* values care has to be taken for other than the rock forming minerals to become dominating the abundance of the given element. This is preferentially true for Ba, Ti, Cr, Co, Ni, Cu and Zn and for element pairs with extreme high or low D_1/D_2 values.

4.3. The characteristics of the distribution of the oxygen isotopes, the alkali, alkaline earth and rare earth elements and the transition elements of the first period between molten natural aluminosilicates and their solidification products can be understood in terms of a coherent model of the structure of molten aluminosilicates: The degree of linking of (Si, Al)O_4 tetrahedrons and the degree of substitution of Si by Al in the tetrahedral centres is approximately not dependent on the aggregational state but only on the concentration of the main elements. The water in magmas is in the molecular state to a considerable extend. The coefficient S in equations (1) and (5) describes the different degrees of linking of (Si, Al)O_4 tetrahedrons, the coefficient k the different water contents of the two aggregational states. (Perhaps the properties of dry magmas could be approximated by equation (1) and (5) using $k = 0$.) The distribution of cations not in tetrahedral positions between liquid and solid aluminosilicates is partially ruled by the radii of the hydrated and non-hydrated ions the radii of the hydrated ions becoming the more important the higher the water content of the magma or the more silicic the magma is, respectively.

4.4. It is due to the peculiarities of the logarithmic function that equation (1) does not only hold for distribution coefficients of single elements, but also for distribution coefficient ratios D_1/D_2 of element pairs at which k and S have to be substituted by the differences $(k_1 - k_2)$ and $(S_1 - S_2)$ of the coefficients of the single elements in question ($k_{1/2} = k_1 - k_2$ and $S_{1/2} = S_1 - S_2$). This enables us to select element pairs with particular $k_{1/2}$ and $S_{1/2}$ values the most importing of which are those with $|S_{1/2}| \leqq 0{\cdot}7$ (little influence of the chemical composition of the magma on D_1/D_2, particularly if $|k_{1/2}| \geqq 0{\cdot}1$ (investigations on the progress of fractional crystallization) or if $|k| \leqq 0{\cdot}065$ (search for common parent magmas and for other than magmatic processes) and those with D_1/D_2 values strongly deviating from unity (search for long intrusive pathways).

Received August 1, 1984
Accepted in revised form October 1, 1984

Literature

[1] *H. Schütze*, Dissertation B, Leipzig 1983.
[2] *K. Wetzel, H. Schütze*, Zs. Geol. Wiss., Berlin (im Druck)
[3] *K. Wetzel*, Isotopenpraxis **21** (1985) 160
[4] *K. Wetzel*, Chem. Erde **42** (1983) 239
[5] *K. Wetzel*, Zs. Geol. Wiss., Berlin (im Druck)
[6] *K. H. Wedepohl*, Handbook of Geochemistry, Berlin – Heidelberg – New York: Springer 1969, 39, 57-71-E-1-19
[7] *R. Hutchison*, Nature **250** (1974) 556
[8] *A. A. Beus*, Geochimija litosfery. Moskva „Nedra" 1981, p. 311
[9] *S. R. Taylor*, Geochim. Cosmochim. Acta **28** (1964) 1273
[10] *A. B. Ronov, A. A. Jaroshevskij*, Geochimija **12** (1976) 1763
[11] *G. Tischendorf, W. Pälchen*, Z. angew. Geologie **19** (1973) 556
[12] *K. Wetzel*, Zs. Geol. Wiss., Berlin (im Druck)
[13] *H. Gerstenberger, G. Haase, G. Tischendorf, K. Wetzel*, Chem. Erde **43** (1984) 263
[14] *H. Schütze, G. Stiehl* et al., ZfI-Mitt. **73** (1976) 232
[15] *K. Wetzel*, Chem. Erde (im Druck)

Isotopenpraxis **21** (1985) 5, S. 159

Buchbesprechung

Biological Effects of Low-level Radiation
Proc. of an International Symposium on the Effects of Low-level Radiation with Special Regard to Stochastic and Non-Stochastic Effects, Venice, Italy, April 1983

Vienna: IAEA 1983. 682 S., div. Abb. u. Tab. (Proc. Series, STI/PUB/646); karton. ÖS 1340.–

Der vorliegende Band aus der Reihe der Tagungsberichte der IAEA enthält die Texte von 46 Vorträgen und 37 Postern. Die Vorträge sind folgenden Sektionen zugeordnet: Epidemiologische Untersuchungen an Überlebenden der Atombombenabwürfe auf Japan; Grundlagenforschung; Genetische Effekte; Epidemiologische Untersuchungen nach medizinischer oder umweltbedingter Strahlenbelastung; Epidemiologische Untersuchungen an beruflich strahlenexponierten Personen; Nicht-stochastische Effekte; Stochastische Effekte; Anwendung von Prinzipien.

Unter dem letzten Titel sind drei Vorträge französischer Autoren über die Anwendung mathematischer Modelle zur Analyse von Dosis-Wirkungs-Beziehungen zusammengefaßt. Eine eindeutige Entscheidung darüber, ob eine lineare Dosis-Wirkungs-Beziehung besteht, bei niedrigen Dosen Reparaturprozesse einen Schwellenwert bewirken (linear-quadratische Beziehung) oder ob sogar bei niedrigen Dosen ein erhöhtes Risiko besteht (vgl. *Kneale* et. al., England), ist noch immer nicht zu treffen, obwohl die meisten Autoren einer linearen Beziehung zuneigen. Entscheidungen über dieses Problem können nur durch statistische Untersuchungen an großen Kollektiven getroffen werden. Bei den neueren Untersuchungen werden jedoch in stärkerem Maße Faktoren wie Geschlecht, Alter während der Exposition, Ganz- oder Teilkörperbestrahlung, besonders gefährdete Organe, sonstige Umwelteinflüsse u. a. berücksichtigt, die zu Modifizierungen früher erhaltener Dosis-Wirkungs-Beziehungen führen. Mit jeder weiteren Aufgliederung entsteht hierbei das Dilemma, daß die zugeordneten Kollektive kleiner und damit statistische Aussagen mit höheren Fehlern behaftet werden. Hinzu kommt, daß zur Zeit alle Dosis-Wirkungs-Beziehungen für die Atombombenopfer von Hiroshima und Nagasaki neu berechnet werden, da die Gammastrahlen- und Neutronenemissionsspektren für die beiden Bombenabwürfe auf Grund neuer Kalkulationen korrigiert werden (*Maruyama*, Japan).

Insgesamt geben die vorliegenden Arbeiten einen guten Einblick in den derzeitigen Stand der Diskussion über die Wirkung kleiner Dosen ionisierender Strahlung, wobei die Beiträge von *P. Metalli*, Italien (Cytogenetische Effekte kleiner Dosen ionisierender Strahlung), von *G. P. Hanson* (USA) und *E. Komarov* (WHO, Genf) über gesundheitliche Wirkungen in Wohngebieten mit hohem natürlichem Strahlenpegel, sowie von *G. Silini*, UNSCEAR Wien, (Überblick über nicht-stochastische Strahlenwirkungen) besonders hervorgehoben seien.

Die zusammenfassenden Darstellungen der Diskussionsschwerpunkte am Schluß der einzelnen Sektionen sind sicher manchem Leser bei der Bewertung der Beiträge hilfreich. Die Vorsitzenden der Sektionen sollten bei zukünftigen Tagungen diesem Aspekt noch mehr Rechnung tragen.

Die Lektüre des Buches wird insbesondere Röntgenologen und Ärzten, die beruflich strahlenexponierte Personen zu betreuen haben, empfohlen. *G. Hübner*

Isotopenpraxis **21** (1985) 5, S. 160 bis 165

Contributions of Isotope Research to the Knowledge of the Distribution Coefficients of Rare Earth Elements between Natural Aluminosilicates and their Molten Phases

K. Wetzel

(Central Institute of Isotope and Radiation Research of the Academy of Sciences of G.D.R.)[1]

The distribution coefficients D of yttrium and the lanthanides between solid and liquid natural aluminosilicates can be described in terms of an extension of Schuetze's [1] incremental method for calculating oxygen isotope effects in silicate systems: $\log D = k + S \cdot \Delta I$, *where* $D = \dfrac{\text{concentration in the solid phase}}{\text{concentration in the liquid phase}}$, ΔI *the difference of the* ^{18}O *indices of the liquid and the solid phase and k and S coefficients to be determined empirically. The coefficient k accounts for the effect of the aggregational state, particularly the water content, S accounts for the effect of the degree of linking of the* $(Si, Al)O_4$ *tetrahedrons in the phases being in equilibrium on the distribution coefficients. The dependence of D, k and S on the atomic number can be understood in terms of the radii of the hydrated and not hydrated lanthanide ions. The constraints on the structural properties of magmas are discussed.*

Die Verteilungskoeffizienten D des Yttriums und der Lanthaniden zwischen festen und flüssigen Alumosilikaten können beschrieben werden auf der Basis einer Erweiterung der Inkrementenmethode von Schütze [1] für die Berechnung von Sauerstoff-Isotopieeffekten in Silikatsystemen: $\log D = k + S \cdot \Delta I$, *worin* $D = \dfrac{\text{Konzentration in der festen Phase}}{\text{Konzentration in der flüssigen Phase}}$, ΔI *die Differenz der* ^{18}O*-Indizes der flüssigen und der festen Phase und k und S empirisch zu ermittelnde Koeffizienten sind. Der Koeffizient k trägt dem Einfluß des Aggregatzustandes, insbesondere des Wassergehaltes, S dem Einfluß des Grades der Vernetzung der* $(Si, Al)O_4$*-Tetraeder in den im Gleichgewicht befindlichen Phasen auf die Verteilungskoeffizienten Rechnung. Die Abhäniggkeit von D, k und S von der Ordnungszahl kann auf der Basis der Radien der hydratisierten und nichthydratisierten Ionen der Seltenerdelemente verstanden werden. Schlußfolgerungen auf die strukturellen Eigenschaften von Magmen werden diskutiert.*

Keywords

equilibrium; isotope effects; isotope ratio; liquids; oxygen 18; rare earths; solids; theoretical data

Aluminosilicate rocks are distinguished by an extensive homogenity and continuity of their structure and chemical composition. $(Si, Al)O_4$ tetrahedrons are the fundamental constituents of all aluminosilicates. The homogenity of structure and chemical composition has its origin in the fact that the bond strengths between oxygen on one hand and silicon or aluminium with coordination number four (Al(IV)) on the other hand predominate over the bond strengths between oxygen and all the other cations. In addition to that O^{2-} is by far the most abundant ion in all aluminosilicates, as well as the ions O^{2-}, Si^{4+} and Al^{3+} forming the $(Si, Al)O_4$ tetrahedrons predominate over all the other ions. Important factors of the continuity of structure and chemical composition of aluminosilicates are the wide abundance of diadochism and the progressive linking of $(Si, Al)O_4$ tetrahedrons which gives rise to the formation of one dimensional chains, two dimensional layers and finally three dimensional networks. Within the course of this linking part of the Si in the centres of the tetrahedrons is progressively substituted by Al.

Thus we can assume that cations other than Si and Al(IV) are situated in a lattice of O^{2-} ions the crystal chemical properties of which are predominantly determined by the degree of linking of the $(Si, Al)O_4$ tetrahedrons and by the extend of substitution of Si by Al in their centres. Thus all the cations not in these centres are surrounded by (4 up to 12) O^{2-} ions in the first coordination sphere and by Si or Al(IV) in a second coordination sphere. Therefore the chemical potentials of cations not in tetrahedral positions should approximately not be dependent on each other.

These structural properties of natural aluminosilicates suggest the extension of *Schuetze*'s ^{18}O indices for calculating oxygen isotopic equilibria between silicatic minerals and rocks [1, 2] also on the calculation of the distribution coefficients

$$D = \frac{\text{concentration in the solid phase}}{\text{concentration in the liquid phase}} \tag{1}$$

of chemical elements between magmas and their crystallization products [2, 3].

The ^{18}O index of an aluminosilicate depends on the chemical composition, particularly on the concentration of O, Si and Al, the valency states, the coordination numbers and the ionic radii of the chemical elements constituting the aluminosilicate and can be approximately calculated by the equation

$$I = \frac{1}{y}(0{\cdot}977x + 0{\cdot}645z + 0{\cdot}0155w + 4{\cdot}092) \tag{2}$$

[1]) Address: DDR-7050 Leipzig, Permoserstraße 15

Tab. 1. *Clarke* values in ppm of Y and the lanthanides in the most abundant plutonic rocks and the concentrations of these elements in the molten phases S_0, S_1, S_2, S_3 and S_4 from which these rocks crystallize, calculated on the basis of the described model. (Figures in brackets are estimated values for lack of a sufficiently great number of experimental data.)

	Y	La	Ce	Pr	Nd	Sm	Eu	Gd	Tb	Dy	Ho	Er	Tm	Yb	Lu
S_0	35·8	34·8	69·3	8·31	35·2	6·90	1·44	7·03	(1·1)	(5·66)	1·46	3·99	0·641	3·83	0·628
basalts (instead of gabbros)	32	6·1	16	2·7	14	4·3	1·5	6·2	1·1	5·9	1·4	3·6	0·60	3·2	0·55
S_1	36·1	37·3	73·9	8·97	37·0										
intermediate rocks	35	31	60	7·4	31	7·12	1·43	7·10	(1·1)	(5·63)	1·47	4·02	0·644	3·88	0·635
(instead of diorites)						6·2	1·3	6·8	1·1	6·1	1·5	3·9	0·65	3·8	0·62
S_2	37·5	45·9	92·9	11·1	45·1	8·38	1·60	7·52	(1·10)	(5)	1·44	4·18	0·636	4·00	0·655
quartz diorites	31	47	97	10·0	45	8·3	2·4	7·6	1·03	(5)	1·21	3·4	0·50	3·0	0·48
S_3	38·2	46·9	92·5	11·2	45·1	8·39	1·50	7·51	(1·10)	(5)	1·46	4·26	0·651	4·11	0·674
granodiorites	33·5	43	83	11·0	44	8·5	2·0	7·4	(1·08)	(5)	1·27	3·7	0·54	3·25	0·54
S_4	42	50	100	11·4	46	8·3	1·10	7·6	1·12	(5)	1·62	4·7	0·74	4·8	0·78
granites	42	50	100	11·4	46	8·3	1·10	7·6	1·12	(5)	1·62	4·7	0·74	4·8	0·78

Tab. 2. Partition coefficients $D_{i/j}$ calculated from *Clarke* values of igneous rocks, coefficients k and S in equation (5), regression coefficients r, per cent deviation $\overline{\Delta}_{i/j}$ of $D_{i/j}$ calculated by equation (5) from $D_{i/j}$ calculated using *Clarke* values and mean per cent deviation Δ of the $D_{i/j}$ values from each other. (Figures in brackets denote $D_{i/j}$ values the evaluation of which includes estimated *Clarke* values or k and S values calculated from those $D_{i/j}$ values, respectively.)

element	$D_{0/1}$ ($\Delta I = 0{\cdot}091$)	$D_{1/2}$ ($\Delta I = 0{\cdot}074$)	$D_{2/3}$ ($\Delta I = 0{\cdot}0215$)	$\Delta_{3/4}$ ($\Delta I = 0{\cdot}019$)	k	S	r	$\Delta_{0/1}$	$D_{1/2}$	$\Delta_{2/3}$	$\Delta_{3/4}$	$\overline{\Delta}$
Y	0·890	0·951	0·819	0·835	−0·093	+0·653	+0·81	4·0	5·1	1·8	0·5	2·9
La	0·169	0·745	0·797	0·888	0·106	−7·18	−0·77	68	50	12	5·1	34
Ce	0·223	0·719	1·046	0·862	0·124	−7·01	−0·79	37	44	10	14	26
Pr	0·313	0·737	0·897	0·973	0·097	−5·27	−0·85	32	31	7·3	2·0	18
Nd	0·388	0·755	0·998	0·966	0·100	−4·61	−0·89	23	24	0·2	6·4	13
Sm	0·613	0·800	0·990	1·019	0·061	−2·67	−0·96	7·2	8·8	1·7	0·4	4·5
Eu	1·045	0·858	1·613	1·538	0·253	−3·23	−0·89	13	20	5·4	1·0	9·9
Gd	0·878	0·930	1·011	0·979	0·0126	−0·695	−0·95	1·4	1·7	1·6	2·0	1·5
Tb	1·000	(1·000)	(0·936)	(0·973)	(−0·0126)	(+0·309)	+0·83	0·4	0·7	2·0	2·0	1·3
Dy	1·045	(1·148)	(1·000)	(1·000)	(−0·0072)	(+0·525)	+0·68	5·0	7·0	0·9	0·6	3·4
Ho	0·956	1·031	0·834	0·825	−0·101	+1·14	+0·89	5·3	6·6	0·6	1·0	3·4
Er	0·899	0·951	0·806	0·826	−0·102	+0·786	+0·87	3·8	4·9	2·1	0·8	2·9
Tm	0·934	1·016	0·777	0·776	−0·135	+1·447	+0·90	6·3	7·7	1·3	0·6	4·0
Yb	0·830	1·176	0·740	0·730	−0·155	+1·67	+0·63	20	21	2·7	3·1	12
Lu	0·871	1·191	0·722	0·743	−0·163	+1·925	+0·71	18	20	4·8	0·7	11

where x, y and z are the formula indices of Si, O and not tetrahedrally coordinated Al in a chemical formula $M^n_{w/n}Al_z[Si_xAl_{4-x}O_y]$ of the silicate, written in accordance with the following order: The sum of the formula indices of Si and Al(IV) has to be four and Al(IV) is written within angular brackets (together with Si and O), n denotes the mean electrical charge of cations not in tetrahedral coordination and $w = 2y - x - 3z - 12$ their mean formula index [4].

As *Schütze* [5] has shown the distribution coefficients of the oxygen isotopes between two aluminosilicates in the liquid or in the solid state with the ^{18}O indices I_1 and I_2 can be calculated by the relation

$$10^3 \ln \alpha = \frac{7\cdot 98 \cdot 10^6}{T^2}(I_2 - I_1) = \frac{7\cdot 98 \cdot 10^6}{T^2}\Delta I \qquad (3)$$

where T is the temperature in K.

With $\Delta I = I_2 - I_1 = 0$ α in equation (3) becomes equal to unity. That is to say the fractionation of the isotopes by crystallization of aluminosilicates is approximately entirely due to the differenc ein the chemical structure and composition between liquid and solid phase. The mere change of the states of aggregation does nearly not induce isotopic fractionation.

Thus the ^{18}O index of an aluminosilicate is a measure of the geometric configuration and the bond relationships of that oxygen ions the lattice of which hosts the cations not in tetrahedral positions that is to say those cations the distribution coefficients of which are investigated in this paper.

Contrary to isotopic fractionation coefficients distribution coefficients of chemical elements will not become equal to unity if there is no difference in the degree of linking of (Si, Al)O_4 tetrahedrons between the liquid and the solid phase. Thus equation (3) has to be extended at least by an additional term k which accounts for the nonequal distribution in the case of equal degree of linking in the two phases:

$$\log D = k + S \cdot \Delta I \qquad (4)$$

where $\Delta I = I_l - I_s$ is the difference of the ^{18}O indices of the magma (I_l) and its crystallization product (I_s) and k and S are coefficients to be determined empirically for each element. In the case of $k = 0$ equation (4) becomes identical with *Schuetze*'s equation[2]) for calculating isotopic equilibria:

$$\log D = S \cdot \Delta I \qquad (5)$$

Equation (5) reveals the physical meaning of the coefficients k and S. S accounts for the influence of the difference in the degree of linking of the (Si, Al)O_4 tetrahedrons with one another and the degree of substitution of Si by Al in the tetrahedral centres between the two phases, while the coefficient k expresses the effect of the mere change of the state of aggregation on the distribution coefficients.

If there is no difference in the degree of linking and the degree of substitution of Si by Al between the two phases ($S = 0$), D is not equal to unity ($\log D \neq 1$) as in the case of isotopic equilibria, but equal to k in accordance with equation (4). That means that there is a difference between liquid and solid phase other than the degree of linking and the degree of substitution of Si by Al which gives rise to deviations from equal partition of chemical elements between the two phases. (It will be shown later that this difference probably is the difference in the water contents of the two phases.)

As we have shown earlier [2], the coefficients k and S can be determined using the *Clarke* values of igneous rocks like gabbros, diorites, quartz diorites, granodiorites and granites, on the basis of the following assumptions:

1. The chemical composition of igneous rocks is mainly determined by fractional crystallization of magmas (and not by anatexis or by assimilation of adjacent rocks).
2. This fractionation proceeds under equilibrium conditions in closed systems, the mean chemical composition of the magmas ascending from the upper mantle region into the continental crust being identical with the mean chemical composition of the continents.
3. The fractional crystallization of these magmas, either continuously or interupted by single other geological processes without differentiating power or of more or less isochemical character, gives rise to the formation of all the igneous rocks in those quantitative proportions which we find in the recent continental crust.

It is the aim of the present paper, to extend these calculations of D, k and S in equation (4) which have so far been restricted to the alkaline elements, the alkaline earth elements and the transition elements of the first period (Sc – Zn), to the rare earth elements and to discuss the results in terms of the geochemical and crystal chemical properties of these elements and in terms of the structure of molten natural aluminosilicates.

Tab. 1 shows the *Clarke* values of Y and the lanthanides in the igneous rocks formed by fractional crystallization and the concentrations of these elements in the molten phases S_0, S_1, S_2, S_3 and S_4, the different types of rocks crystallize from in the following proportions: 7·9 weight-% gabbros, 53·1 weight-% diorites (of the lower part of continental crust), 3·8 weight-% quartz diorites, 15·6 weight-% granodiorites out 19·6 granites and quartz monzonites [4].

Tab. 2 shows the distribution coefficients $D_{i/j}$ of the rare earth elements between the different rocks and the molten phases S_i or S_j, respectively, resulting before and after solidification of rock type i/j:

$$D_{i/j} = \frac{c_{i/j}}{\frac{1}{2}(c_i + c_j)}$$

where c are element concentrations. Regression calculations on the basis of equation (5) using these $D_{i/j}$ values yields the coefficients k and S and the regression coefficients r in Tab. 2.

Discussion

As Tab. 2 indicates 10 of 15 rare earth elements show regression coefficients greater than 0·80. In 9 of 15 cases the mean per cent deviation $\bar{\Delta}$ of the $D_{i/j}$ values calculated by equation (5) from $D_{i/j}$ calculated using Clarke values does not exceed 10%, the mean $\bar{\Delta}$ value being 9·9%. Thus equation (5) describes the influence of the chemical composition on the distribution coefficients of rare earth elements between magmas and their solidification products rather adequately.

Nevertheless comparison with data calculated on independent ways would be highly appreciable. Involvment of data about volcanic rocks, the solidification conditions of which deviate drasticly from those of plutonic rocks in-

---

[2]) In the case of oxygen isotope distribution coefficients in aluminosilicate systems S in equation (4) has to be substituted by $\frac{3,47 \cdot 10^3}{T^2}$, where T is the temperature in K.

Tab. 3. *Clarke* values in ppm of Y and some rare earth elements in andesites and rhyolites and the concentrations of these elements in the molten phases S_0 and S_1 from which these rocks are supposed to crystallize, distribution coefficients $D_{0/1}^{exp.}$ calculated from *Clarke* values, distribution coefficients $D_{0/1}^{calc.}$ calculated by equation (5) using k and S values in Tab. 2 and per cent deviation Δ of $D_{0/1}^{calc.}$ from $D_{0/1}^{exp.}$. (Elements with *Clarke* values less than 1 ppm are omitted for lack of data or because of their uncertainty, respectively.)

	Y	La	Ce	Pr	Nd	Sm	Eu	Gd	Dy	Er	Yb
S_0	21·5	22	30·3	4·23	17·3	3·84	1·0	4·19	4·0	2·67	2·67
andesites	19·7	19·5	29·2	3·6	16·9	3·7	1·0	4·1	3·90	2·5	2·5
S_1	36	42	38·9	9·3	20·3	5	1·0	4·9	5	4·0	4·0
rhyolites	36	42	38·9	9·3	20·3	5	1·0	4·9	5	4·0	4·0
$D_{0/1}^{exp.}$	0·685	0·609	0·844	0·532	0·899	0·837	1·00	0·837	0·867	0·749	0·750
$D_{0/1}^{calc.}$	0·892	0·425	0·455	0·558	0·621	0·765	1·09	0·926	0·833	0·892	0·904
Δ	30	43	30	4·9	31	8·6	9	11	3·9	19	21

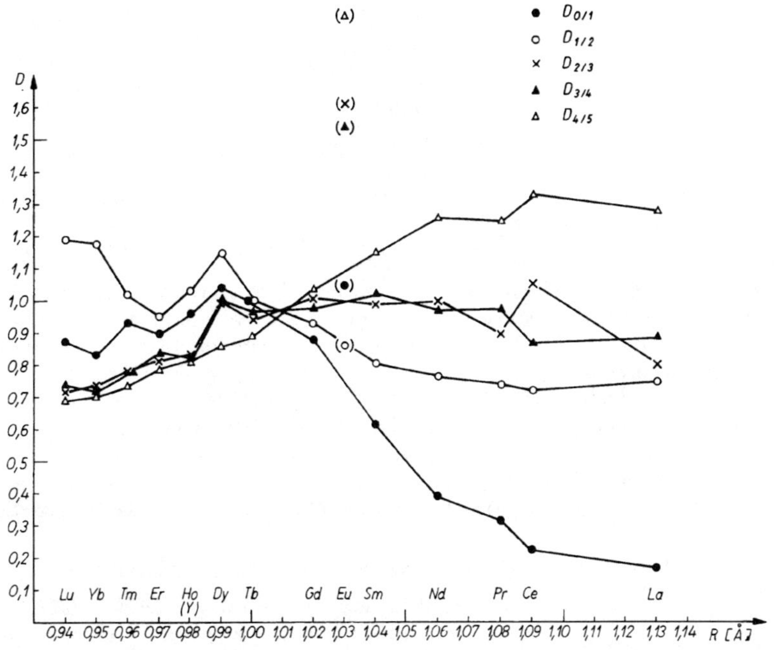

Fig. 1. Dependence of the distribution coefficients $D_{i/j}$ on the ionic radii R (coordination number VI) of the trivalent cations of rare earth elements

vestigated so far, would be of particular interest. This can be done in the following way. As *Daly* [6] has shown, andesites and rhyolites amount to 16 or 2%, respectively, of the entire area of continents covered by volcanic rocks. The bulk of these rocks is generated in subduction zones being formed by fractional crystallization from common parent magmas having ascended from mantle region. (As Sm—Nd and other geochemical investigations have shown the contamination of these magmas by crystal adjacent rocks is restricted to a small extend.) If we assume that the amounts of andesites and rhyolites are proportional to the areas they cover, we get the scheme shown in Tab. 3, using *Clarke* values by *Cole* et al. [7], *Taylor* [8], *Bokhari* [9], *Ewart* et al. [10] and *Wedepohl* [11].

As Tab. 3 shows there is a rather good agreement between the distribution coefficients calculated from *Clarke* values and those calculated using equation (5) with k and S values taken from Tab. 2, the mean relative deviation being 19% which again gives evidence of the validity of equation (5) and of the limited effect of solidification conditions on the distribution coefficients.

Fig. 1 demonstrates the distribution coefficients $D_{i/j}$ calculated from *Clarke* values as a function of the ionic radii R of the trivalent cations.

Apart from Eu which occurs in trivalent and divalent state in nature there is a clear coherence between $D_{i/j}$ and R even in the case of yttrium which does not belong to the 4 f elements. Nevertheless the distribution coefficients for basic and intermediary rocks decrease with increasing R, while the distribution coefficients for quartz diorites and granodiorites show a maximum near samarium which evidently cannot be interpreted in terms of the R values alone. It is necessary to assume an effect counteracting the influence of R. This constraint is supported by the slope of the curve for the distribution coefficient $D_{4/5}$ for granites in Fig. 1 which can be calculated by equation (5) using $\Delta I = 0$: $D_{4/5}$ does not decrease, but increases with increasing R. Thus $D_{0/1}$ and $D_{1/2}$ on one hand and $D_{4/5}$ represent opposite types with respect to their dependence on R the $D_{1/2}$ and $D_{2/3}$ lying between these types in the following sense: The light rare earth elements behave like the $D_{0/1}$ and $D_{1/2}$ type and the heavy ones like the $D_{4/5}$ type the distribution coefficients of Y following the D values of the heavy rare earth elements.

The $D_{0/1}$ and $D_{1/2}$ type behaviour can easily by understood: The heavier lanthanides with their smaller ionic radii R are enriched in the solid phase with its smaller degree of linking and its higher density the larger ions

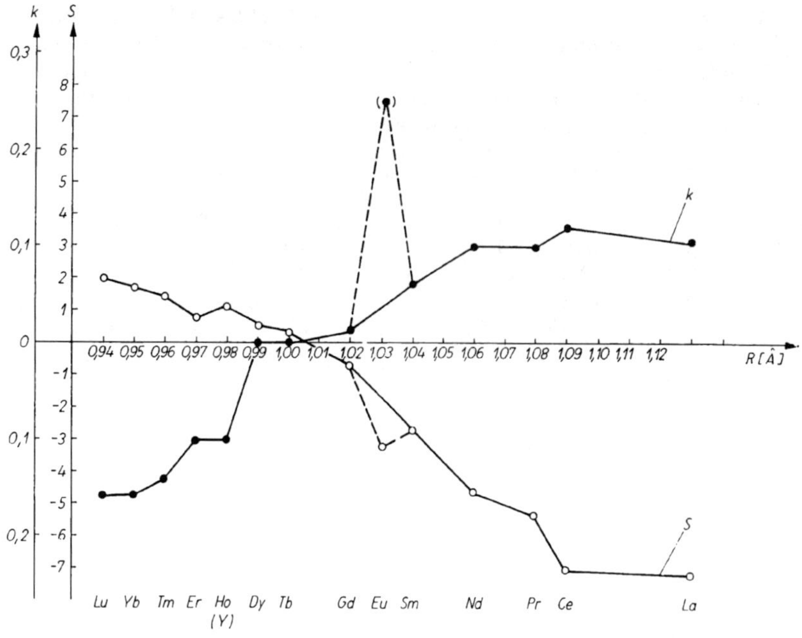

Fig. 2. Dependence of the coefficients k and S on the ionic radii R of the rare earth ions

prevailing in the liquid phase distinguished by a higher degree of linking and therefore a smaller density. The most obvious way to understand the dependence of the distribution coefficients $D_{4/5}$ on R is to take into account the generally higher water content of silicic magmas in comparison with basic and intermediate magmas. The higher the water content of a magma the more pronounciated becomes the formation of hydrated ions, particularly hydrated cations. As is well known from investigations on aqueous solutions of rare earth compounds the radii of hydrated ions increase with decreasing radii of the non hydrated ions. Thus in silicic magmas with their high water contents the distribution of the lanthanides between liquid and solid phase is ruled by the sequence of the radii of hydrated ions which is opposite to the sequence of the radii of not hydrated ions. The discussion of the k and S values and their dependence on the ionic radii will have to prove whether this interpretation is correct or not.

Fig. 2 shows the coefficients k and S in equation (5) as a function of the ionic radii R. As we have seen, the coefficient S describes the influence of the degree of linking of the (Si, Al)O_4 tetrahedrons. (If there is no effect of the difference in aggregational state on D, the coefficient k becomes zero and $\log D = S \cdot \Delta I$ according to equation (5).) As is seen in Fig. 2, S shows the behaviour to be expected: The small ions of the heavy rare earth elements prefer the solid phase with its lower degree of linking and higher density the lighter lanthanides prevailing in the liquid phase distinguished by a high degree of linking and lower density. On the other hand the coefficients k behave in the opposite manner: Apart from the usual Eu anomaly k increases with increasing ionic radius R. This inversion again can be attributed to the inversion of the sequence of ionic radii owing to hydration. If the effect of the degree of linking recedes ($S = 0: \log D = k$), the light lanthanides with their smaller hydrated ions prevail in the solid phase, the larger hydrated ions of the heavy rare earth elements being enriched in the liquid phase with its higher water content and its lower density.

Thus we can state: If the situation is ruled by the sequence of the radii of not hydrated lanthanide ions the correlation between the distribution coefficients and the ionic radii is negative, if the situation is ruled by the sequence of the radii of the hydrated ions, this correlation becomes a positive one. Considering these clear correlations between D, k, S on one hand and R on the other hand it should be allowed to propose corrected coefficients k and S for practical use: Linear regression using the data on Tab. 2 with the exception of those regarding Eu (because of its anomalous behaviour), Tb and Dy (because of some uncertainties in their *Clarke* values) yields the equations

$$k = -1.80 + 1.75 \cdot R$$

$$S = +54.28 - 55.01 \cdot R$$

with correlation coefficients of $r = 0.96$ or -0.98, respectively. Using these equations we get the coefficients k^x and S^x given in Tab. 4.

Tab. 4. Coefficients k^x and S^x corrected by fitting k and S values in Tab. 2 (Eu values are excluded from fitting.)

	R	k	S
Y	0.98	−0.085	+0.366
La	1.13	+0.178	−7.89
Ce	1.09	+0.108	−5.69
Pr	1.08	+0.090	−5.14
Nd	1.06	+0.055	−4.04
Sm	1.04	+0.020	−2.93
Eu	1.03	+0.253	−3.23
Gd	1.02	−0.0149	−1.83
Tb	1.00	−0.050	−0.734
Dy	0.99	−0.067	−0.184
Ho	0.98	−0.085	+0.366
Er	0.97	−0.102	+0.916
Tm	0.96	−0.120	+1.466
Yb	0.95	−0.137	+2.016
Lu	0.94	−0.155	+2.567

Conclusions

Equation (5) can be used for the approximative calculation of distribution coefficients between solid and liquid aluminosilicates not only in the case of the alkaline and alkaline earth elements and the transition elements of the first period, but also in the case of Y and the lanthanides. The coefficients k and S in this equation and their dependence on the radii of the hydrated and not hydrated ions of the rare earth elements can be understood on the basis of the following model of the structural properties of magmas: The degree of linking of $(Si, Al)O_4$ tetrahedrons and the degree of substitution of Si by Al in tetrahedral positions is more dependent on the chemical composition of the aluminosilicate system than on the aggregational state. The water in magmas to a considerable extend exists in the form of water molecules and not only in the form of hydroxy groups. (The latter conclusion is strongly supported also by investigations on the isotopic composition of oxygen in rocks and minerals [4].)

Received July 27, 1984

Literature

[1] *H. Schütze*, Chem. Erde 39 (1980) 321

[2] *K. Wetzel, H. Schütze*, Zs. Geol. Wiss. (Berlin) (im Druck)

[3] *K. Wetzel*, Zs. Geol. Wiss. (Berlin) (im Druck).

[4] *K. Wetzel*, Chem. Erde 42 (1983) 239

[5] *H. Schütze*, Dissertation B, Leipzig 1983

[6] *R. A. Daly*, Igneous Rocks and the Depth of the Earth. New York: McGraw-Hill Book Co. 1933

[7] *J. W. Cole, K. V. Cashman, P. C. Ranking*, Chem. Geol. 38 (1983) 255

[8] *S. R. Taylor*, Island Arc Models and the Composition of the Continental Crust, in Island Res., Deep Sea Trenches and Back Arc Basins (eds. *M. Talwane, W. C. Pitman*) Maurice Erving Ser. I, American Geophysical Union, Washington 1977, 325

[9] *F. Y. Bokhari, J. D. Kramers*, Earth Planet. Sci. Lett. 54 (1981) 409

[10] *A. Ewart, W. B. Bryan, J. B. Gill*, J. Petrol. 14 (1973) 429

[11] *K. H. Wedepohl*, Handbook of Geochemistry. Berlin – Heidelberg – New York: Springer, Vol. II/5,39/57-71-E-15

Isotopenpraxis 21 (1985) 5, S. 165 bis 169

Isotopenzusammensetzungen von Strontium und Neodym als Kennzeichen von Erdmantelmaterial*)

S. B. Brandt, V. S. Lepin, M. N. Maslovskaja

(Institut für die Erforschung der Erdkurste, Irkutsk)[1])

Es wird gezeigt, daß der Anteil von Erdmantel-, Erdkrusten- und sedimentären Material in Gesteinen und Erzlagerstätten mit Hilfe von Isotopenmethoden ermittelt werden kann. Am Beispiel von jakutischen Kimberliten wurden Mischungsprozesse von Erdmantel- und Krustenmaterial mit Hilfe der Strontiumisotope nachgewiesen. Wegen der hohen Empfindlichkeit des Strontiums gegenüber hydrothermalen Einflüssen wird die kombinierte Untersuchung der Neodym-Strontium-Isotope als geeigneter angesehen, das Problem der Ermittlung des Anteils des Erdmantelmaterials zu klären. Dies wird am Beispiel von Seltenterd-Mineralisationen und Alkaligesteinen Ostsibiriens und der Mongolischen Volksrepublik gezeigt.

It is shown that the share of mantle, crustal and sedimentary substances in rocks and ore formations may be determined by isotope methods. On an example of Yakutian kimberlites a prove of mixing processes of mantle and crustal substances by means of strontium isotopes is delivered. Due to the high sensitivity of strontium to hydrothermal disturbances, the combined use of neodymium and strontium isotopes is believed to be a more perfect approach to the problem of determining the share of mantle substances. This is demonstrated by an application of isotopes to REE-minerals and alkaline rocks of Eastern Siberia and the Peoples Republic of Mongolia.

Keywords

earth crust; geochemistry; hafnium isotopes; isotope ratios; minerals; neodymium isotopes; ores; rare earths; strontium isotopes

Einleitung

Mantelmaterial ist der unmittelbaren Beobachtung entzogen. Schlußfolgerungen über den Charakter von Mantelmaterial beruhen auf seismischen Daten, thermodynamischen Berechnungen subkrustaler Mineralparagenesen, auf der Geochemie einiger Elemente und schließlich auf Bilanzen der Radioaktivität und der Isotopenbestände. In der modernen Geologie stehen Probleme der Genese von Mineralen und Gesteinen im Vordergrund. Die Plattentektonik ist der kontinuierlichen Genese der geologischen Formationen gewidmet. Selbstverständlich ist die Frage über den Anteil von Krusten-, Mantel- und Sedimentmaterial in den Begriff „Genese“ mit eingeschlossen. Diese Frage ist auch von großer praktischer Bedeutung. Das sei an einigen Beispielen erläutert. Betrachten wir zunächst das Problem der Genese der Urtite (Rohstoff für die Aluminiummetallurgie). Zwei Annahmen sind möglich: die Existenz von Urtitmagmen in großer Tiefe einerseits und die Bildung von Urtiten im Kontakt mit Marmor andererseits. Jeder Standpunkt bedingt völlig verschiedene Strategien der Suche und Prospektion und damit unterschiedliche Kosten. Dasselbe würde auch für die Prospektion von primären Diamantlagerstätten zutreffen. Sind diamantführende Kimberlite Mantelursprungs, so müßte die Prospektion auf Systeme von Tiefenbrüchen orientiert sein. Sind aber Kimberlite metasomatischer Herkunft, dann müssen grundsätzlich andere Suchkriterien gelten. Chrom, Nickel und Metalle der Platingruppe können nur mit Gesteinen von Mantelherkunft verknüpft sein usw.

*) Vortrag, gehalten auf der 3. Arbeitstagung „Isotope in der Natur“, 14.–18. 11. 1983 in Leipzig, DDR

[1]) Anschrift: SU-664033, Irkutsk 33, Lermontowa 128, UdSSR

In der letzten Zeit haben die Isotopenverhältnisse von solchen chalkophilen Elementen wie Strontium, Neodym und Hafnium als Kriterien für eine Mantelherkunft mannigfaltige Anwendungen gefunden.

Strontiumisotope

Strontium ist in der Natur weit verbreitet, bildet keine flüchtigen Verbindungen und tendiert zur Akkumulation in der Hydrosphäre. Für das Strontium sind einige Bezugswerte der Isotopenverhältnisse kennzeichnend.

1. Zur Zeit der Nukleosynthese hat ein gewisser Mindest-(Initial)-Gehalt des radiogenen Isotops ^{87}Sr im Gesamtstrontium Platz gehabt. Das diesem Zeitpunkt entsprechende $^{87}Sr/^{86}Sr$-Verhältnis von 0,698 ... wird als BABI (Basalt Achondrite Best Initial) oder ADOR (nach dem Meteoriten Angre dos Reis) bezeichnet und wurde an Meteoritproben bestimmt. Dieser Wert wird als charakteristisch für das ursprüngliche Erdmantelmaterial angenommen. Im gegenwärtigen Mantel ist er infolge der Ansammlung des radiogenen Isotops Strontium-87 etwas höher (0,7020 – 0,7040).
2. Im Weltozean ist die gegenwärtige Isotopenzusammensetzung des Strontiums erstaunlich homogen ($^{87}Sr/^{86}Sr = 0{,}7093$). Das Strontium in Sedimentgesteinen hat eine Isotopenzusammensetzung, die nahe diesem Wert liegt.
3. Rubidiumhaltige Krustengesteine sind an radiogenem Strontium angereichert.

Bei der Untersuchung der Strontiumsiotopenverteilung gewonnene Erfahrungen deuten auf weit verbreitete Mischungsprozesse hin. Es läßt sich zeigen [1], daß in einer Mischung zweier Substanzen (1 und 2) die Isotopenzusammensetzung des Strontiums durch eine hyperbolische Abhängigkeit vom Strontiumgehalt gekennzeichnet ist:

$$\frac{b_1 - b}{b - b_2} = \beta \frac{c_1 - c}{c - c_2} \tag{1}$$

($b = {}^{87}Sr/^{86}Sr$; c = der Gehalt eines Elements, z.B. des Strontiums, $\beta = {}^{86}Sr_2/^{86}Sr_1$).

Gleichung (1) kann leicht in folgende Form gebracht werden:

$$^{87}Sr/^{86}Sr = A(C + A/B) + G/B$$

Da $C \leqq A/B$, ist

$$^{87}Sr/^{86}Sr = A/Sr + G/B \tag{2}$$

(A, B, G sind Funktionen von $b_{1,2}$; $c_{1,2}$).

Mischungsprozesse von Strontiumisotopen werden durch eine Gerade im $^{87}Sr/^{86}Sr$-1/Sr-Diagramm dargestellt.

Betrachten wir nun die Verteilung der Strontiumisotope in Karbonaten aus Kimberliten Jakutiens [2, 3]. Die $^{87}Sr/^{86}Sr$-Meßwerte für Kimberlitgesteine und deren Minerale liegen zwischen 0,7030 und 0,7154. Der Mindestwert entspricht sicher Mantelbildungen, der Höchstwert sialischen Krustenbildungen. Eine Reihe von dazwischen liegenden Werten ist ebenfalls gemessen worden. Es entsteht die Frage, ob in diesem Falle eine kontinuierliche Mischungsreihe von Mantel- und Krustenmaterial vorliegt oder nur ein Einfluß der sedimentären Nebengesteine des Sibirischen Tafellandes (Kalksteine und Argillite). Diese Frage kann nur durch Heranziehen der Gleichung (2) entschieden werden.

In Abb. 1 ist die Beziehung zwischen $^{87}Sr/^{86}Sr$ und 1/Sr dargestellt. Als Bezugspunkt für die Mantelzusammensetzung wurde ein Karbonatit aus der Umgebung der

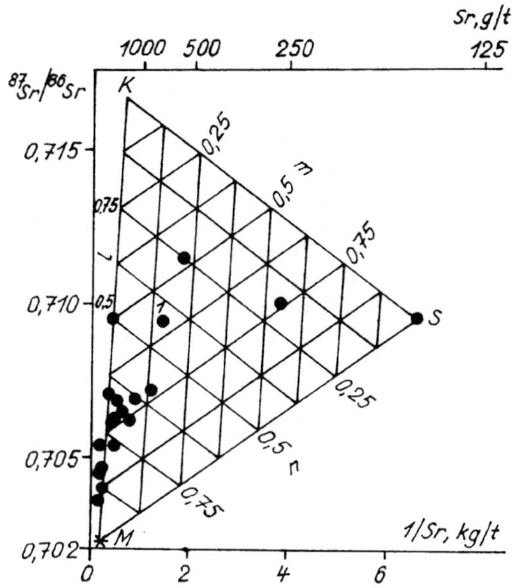

Abb. 1. Dreikomponentendiagramm der Mischungsprozesse in Kimberliten Jakutiens

M, K, S – Bezugspunkte für Mantel-, Krustenmaterial und Sedimente

Kimberlitpipe genommen, der ein besonders niedriges Sr-Isotopenverhältnis ($^{87}Sr/^{87}Sr = 0{,}7023$) bei einem extrem hohen Strontiumgehalt von 6,5 kg/t aufwies.

Als Bezugspunkt für sedimentäre Bildungen liegt das karbonatische Nebengestein vor (0,7093; 0,14 kg/t). Als Bezugswerte des silikatischen Krustenmaterials wurden aus Untersuchungen am kristallinen Fundament des Sibirischen Tafellandes ermittelte Werte (0,7160; 1,0 kg/t) gewählt. Da diese Bezugspunkte verbindenden Geraden (M = Mantel; K = Kruste; S = Sedimente) stellen Mischungsprozesse zwischen zwei betreffenden Bezugswerten dar und die von ihnen begrenzte Fläche ein Mischungsdiagramm für drei Komponenten. Es ist zu sehen, daß die Mehrzahl der Punkte in der Nähe der Geraden MK liegt, während die Felder der Mantel-Sediment- und Krusten-Sediment-Mischungen unbesetzt bleiben. Das bedeutet, daß die Isotopenverhältnisse des Strontiums hauptsächlich durch Mischung von Mantel- und Krustenmaterial gebildet werden trotzt der scheinbaren Übereinstimmung einiger Daten mit sedimentären Werten. Nur vereinzelte Proben von Karbonaten des metasomatisch überprägten Stadiums erfahren eine Abweichung in Richtung des sedimentären Bezugspunktes. Das in Abb. 1 dargestellte Mischungsdreieck könnte als ein im $^{87}Sr/^{86}Sr$-1/Sr-Diagramm konstruiertes isotopisch-petrologisches Mischungsdiagramm aufgefaßt werden. Jedem Punkt innerhalb des Dreiecks können ein Sr-Isotopenverhältnis, der Sr-Gehalt und gleichzeitig auch die Anteile der Bezugssubstanzen (Mantel, Kruste, Sedimente) zugeordnet werden auf Grund folgender Beziehungen (siehe [4]):

$$l/C_M + m/C_K + n/C_{Sed} = 1/C_{Sr}$$

$$l + m + n = 1$$

Beispielsweise hat der Punkt 1 (Abb. 1) die Werte $^{87}Sr/^{86}Sr = 0{,}7095$ und $Sr = 0{,}75$ kg/t und repräsentiert eine Mischung aus 45% Mantel-, 42% Krusten- und 13% sedimentärem Material.

Ein fehlender Nebengesteinseinfluß ist auch für südafrikanische Kimberlite in der jüngst veröffentlichten Arbeit von *Exley* und *Jones* [5] festgestellt worden.

Demzufolge können durch die Nutzung der Strontiumisotope in bestimmten Fällen sowohl Mischungsvorgänge der Isotope erkannt als auch hinreichend begründete Schlußfolgerungen über Anteile von Mantel-, Krusten- und sedimentären Substanzen an der Bildung von Gesteinen gezogen werden.

Kombinierte Neodym- und Strontiumisotope

Ein alternatives und vielleicht ein besseres Verfahren zur Bestimmung der Mantel- und Krustenanteile stellt die in der jüngsten Zeit praktizierte Kombination von Neodym und Strontiumisotopenuntersuchungen dar [6–8]. Die Messung der Neodymisotope erhöht erheblich die Sicherheit von Schlußfolgerungen über die Genese.

Betrachten wir die isotopengeochemischen Eigenschaften des Neodyms. Das natürliche Neodym stellt eine Mischung von sieben Isotopen mit den Massenzahlen 142, 143, 144, 145, 146, 148 und 150 dar. Die Menge des Isotops ^{143}Nd in der Natur wächst infolge der Kernreaktion $^{147}\text{Sm} \rightarrow \alpha + {}^{143}\text{Nd}$ mit einer Halbwertszeit von $125 \cdot 10^9$ a. Der relative Anteil des radiogenen Isotops ^{143}Nd wird üblicherweise durch das Verhältnis ^{143}Nd/^{144}Nd angegeben. Zur Eliminierung von Fraktionierungseffekten (sowohl natürlicher als auch meßtechnischer) wird ebenso wie beim Strontium eine Normalisierung durchgeführt. In diesem Falle wird auf ein Verhältnis zweier stabiler Isotope (^{146}Nd/^{144}Nd) zurückgeführt. Über den Standardbezugswert dieses Verhältnisses besteht in den Veröffentlichungen bis jetzt noch keine einheitliche Meinung. Mitarbeiter des California Institute of Technology [8] benutzen den Wert 0,724107, während in der französischen Literatur der Wert 0,721900 bevorzugt wird [9]. Zieht man in Betracht, daß Messungen in einer Reihe führender Laboratorien mit einer Genauigkeit von 10^{-6} bis 10^{-7} ausgeführt werden, so wird deutlich, daß eine Diskrepanz des Normalisierungsverhältnisses bereits in der dritten Stelle alle nachfolgenden Ziffern an Absolutwert ungültig macht. Das Auftreten von systematischen Fehlern schließt jedoch nicht die Möglichkeit aus, Probenserien, Reihenfolgen und Gesetzmäßigkeiten zu analysieren, die bei einem einheitlichen Normalisierungsverfahren nicht gestört werden. Nur Meßwerte verschiedener Laboratorien können nicht miteinander verglichen werden. In der vorliegenden Arbeit wurde die Normalisierung mit beiden Standardwerten durchgeführt.

Das Sm–Nd-System unterscheidet sich in zweierlei Hinsicht grundsätzlich vom Rb–Sr-System.

Erstens: Das Sm/Nd-Verhältnis ist im Mantel höher als in der Kruste. Im Gegensatz dazu ist das Rb/Sr-Verhältnis im Mantel niedriger als in der Kruste. Demzufolge sind Mantel- im Unterschied zu Krustenbildungen an radiogenem Neodym angereichert, während die Erdkruste an radiogenem Strontium angereichert ist. Der mögliche Mindestwert des ^{143}Nd/^{144}Nd-Verhältnisses beträgt 0,505828 [10], welcher wohl dem Anfangszustand der Erde entsprechen dürfte. Die besten Annäherungen an diesen Grenzwert konnten an Gesteinen der ältesten Erdkruste festgestellt werden. Diese Gesetzmäßigkeit ist in dem Evolutionsdiagramm in Abb. 2 dargestellt. Dabei ist das Teildiagramm a das bekannte Evolutionsdiagramm der Strontiumisotope [11], b ein entsprechendes Diagramm für das Neodym. Durch Eliminieren des Zeitparameters erhält man „Bahnkurven" (Teildiagramm c). Auf den Geraden 1–1′ und 2–2′ liegen Meßpunkte, die rezente Mischungen von Mantel- und Krustenmaterial bzw. solche in der geologischen Vergangenheit repräsentieren.

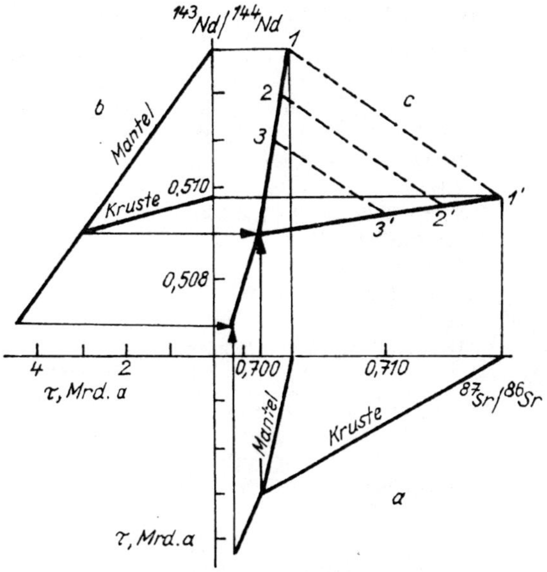

Abb. 2. Evolutionsdiagramm für Strontium- und Neodymisotope
a = Evolutionsdiagramm für Strontiumisotope; b = Evolutionsdiagramm für Neodymisotope; c = Mischungslinie der Strontium- und Neodymisotope, durch Eliminierung des Zeitparameters erstellt

Aus veröffentlichten Daten folgt, daß die Steigung der Geraden 1–1′ (Δ ^{143}Nd/^{144}Nd : Δ ^{87}Sr/^{86}Sr) ca. –0,25 beträgt.

Zweitens: Die geochemischen Eigenschaften der Seltenerd-Elemente der Cergruppe Samarium und Neodym sind sehr ähnlich, während die Eigenschaften des Rubidiums und Strontiums sich wesentlich unterscheiden. Das ist bei hydrothermalen Prozessen von großer Bedeutung. Seltene Erden sind hydrophob und nicht flüchtig. Demzufolge wird die Isotopenzusammensetzung des Neodyms bei metasomatischen Einflüssen nicht verändert. Strontium dagegen geht leicht in die flüssige Phase über und vermischt sich, wobei seine Isotopenzusammensetzung verändert wird. Dieses wurde bei Untersuchungen der Isotopenverhältnisse an Alkalingesteinen von Roccamonfina (Italien) [13], am Ophiolit-Komplex von Samail (Kuweit) und an Graniten verschiedenen Ursprungs festgestellt. In allen genannten Fällen weichen die Meßpunkte im ^{143}Nd/^{144}Nd-^{87}Sr/^{86}Sr-Diagramm von der Regressionsgeraden ab, und zwar so, daß ein und demselben ^{143}Nd/^{144}Nd-Wert eine gewisse Anzahl von ^{87}Sr/^{86}Sr-Werten entspricht. Die bisher veröffentlichten Arbeiten über die Verteilung der Neodym- und Strontiumisotope beziehen sich hauptsächlich auf die Ermittlung grundlegender Isotopenbeziehungen zwischen Kruste, Mantelreservoir und Meteoriten. Sie sind der Konstruktion des Diagramms in Abb. 2 zugrundegelegt worden. Der moderne Stand der Forschungen ermöglicht eine Anwendung der Nd-Sr-Isotope zur Lösung von Problemen der Erzgenese und anderen aktuellen Fragen der angewandten Geologie. In der vorliegenden Arbeit wird sie auf Seltenerd-Mineralisationen und Alkaligesteine Ostsibiriens und der Mongolischen Volksrepublik (MVR) angewandt. Dabei besteht das Hauptinteresse in der Ermittlung der Möglichkeiten dieser Methode bei der Untersuchung der Herkunft des Erzes, des Mechanismus seiner Anreicherung und der Migrationswege.

Es werden 9 Proben aus den Alkaliultrabasit-Massiven Schidoj im Ostsajan und Muschugaj-Chuduk (MVR) sowie des Massivs Chan-Bogdo (MVR) untersucht. Eine geologische Beschreibung sowie die Meßwerte sind der Tabelle

Tab. 1. Beschreibung der Proben und Meßwerte

Probe Nr.	Mineral, Gestein	Massiv	Sr kg/t	Nd kg/t	$^{87}Sr/^{86}Sr$	$^{142}Nd/^{144}Nd$
1	Calcit aus einem Karbonatit	Schidoj	10,0	1,6	0,7042	0,51270
2	Apatitgestein	Muschugaj-Chuduk	24,0	13,0	0,7055	0,51240
3	dito	dito	18,7	23,0	0,7052	0,51240
4	Apatit aus einem Apatit-Sphengestein	Schidoj	8,8	13,7	0,7101	0,51140
5	Sphen (Titanit)	Chan-Bogdo	1,0	6,0	0,7061	0,51226
6	Armstrongit	dito	0,2	0,32	0,7065	0,51225
7	Ankylit	Schidoj	—	—	0,7082	0,51167
8	Seltenerdsilikat	Chan-Bogdo	0,2	30,0	0,7094	0,51264
9	Bastnäsit aus einem Karbonatit	Lutschingol	1,5	52,0	0,7093	0,51228

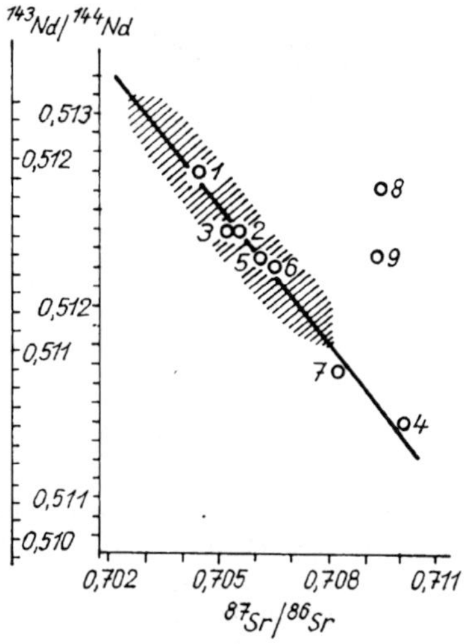

Abb. 3. Beziehung zwischen $^{87}Sr/^{86}Sr$ und $^{143}Nd/^{144}Nd$ für Seltenerdminerale und Alkaligesteine Ostsibiriens und der Mongolischen Volksrepublik

zu entnehmen. Eine ausführlichere geologische Beschreibung und Meßtechnisches sind in unserer Arbeit [12] enthalten. In Abb. 3 sind die Meßergebnisse in der Form eines $^{143}Nd/^{144}Nd$-$^{87}Sr/^{86}Sr$-Diagramms wiedergegeben. Die Mehrzahl der Proben liegt auf oder nahe der bekannten Mischungsgeraden von Mantel- und Krustenmaterial [12]. Der schraffierte Bereich entspricht den für ozeanische Basalte und Inselbögen ermittelten Werten. Demzufolge ist das Verhalten des Neodyms und des Strontiums in dieser Probengruppe im Laufe der Kristallisation der Seltenerd-Minerale streng kongruent. Einem Mantelwert von $^{143}Nd/^{144}Nd$ entspricht immer ein Mantelwert von $^{87}Sr/^{86}Sr$, ausgenommen die Punkte 8 und 9 (Abb. 3). Wie schon gesagt, ist das Strontium im Gegensatz zu Seltenerd-Elementen hydrophil und in hydrothermalen Prozessen mobil. Die Lage der Punkte 8 und 9 wird durch Einwirkung hydrothermaler Prozesse erklärbar. Punkt 8 entspricht einem Seltenerd-Silikat aus einem Pegmatit (Massiv Chan-Bogdo), das eine Umwandlung der Eigenstruktur in einen amorphen Zustand erlitten hat und durch einen jüngeren hydrothermalen Quartz „eingebacken" wurde; Punkt 9 entspricht einem Bastnäsit (Massiv Lutschingol) aus hydrothermalen Bastnäsit-Quarz-Gängen, die durch Verwitterung verändert worden sind. In beiden Fällen fand wahrscheinlich eine Mobilisierung des Mineralsubstrats unter Beteiligung eines mit Strontium ($^{87}Sr/Sr^{86}$ = 0,7093) gesättigten aktiven Fluids statt. Dieser Wert ist für eine bestimmte Klasse von hydrothermalen Mineralen kennzeichnend.

Demnach sind bei kombinierten Neodym-Strontium-Isotopenuntersuchungen auch Abweichungen von der normalen Mischungsgeraden informativ. Diese Methode erscheint als besonders wertvoll für Erz- und petrogenetische Untersuchungen.

In der Logik gibt es die Begriffe „notwendig" und „hinreichend". Ausgehend von der in der letzten Zeit festgestellten Heterogenität des Mantels sowie den soeben beschriebenen Erscheinungen kann man sagen, daß ein mit dem BABI-Wert identisches $^{87}Sr/^{86}Sr$-Verhältnis nicht mehr eine *notwendige* Bedingung für Mantelzugehörigkeit ist, sondern eine *hinreichende.* Gleichzeitig könnte ein hoher $^{143}Nd/^{144}Nd$-Wert als eine gleichzeitig *notwendige und hinreichende* Bedingungen für Mantelmaterial bezeichnet werden.

Eine Möglichkeit kombinierter Strontium-Neodym-Hafnium-Isotopenuntersuchungen

Kürzlich ist es *Patchett* und *Tatsumoto* [14] gelungen, eine Isochrone für eukritische Meteorite mittels der Lutetium-Hafnium-Methode zu erstellen sowie den Wert der β-Zerfallskonstante des Lutetium-176 zu präzisieren. Analog zu Strontium-87 und Neodym-143 wächst der Gehalt des Hafnium-176 in geologischer Zeit. Es wird durch das Isotopenverhältnis $^{176}Hf/^{177}Hf$ charakterisiert. Die Seltenen Erden Lutetium und Hafnium verhalten sich geochemisch wie Samarium und Neodym, wobei allerdings ihr Vorkommen bedeutend seltener ist als das von Neodym und Samarium. Die „Trägheit" der Hafniumisotope in bezug auf metasomatische Einwirkungen ist noch größer als bei

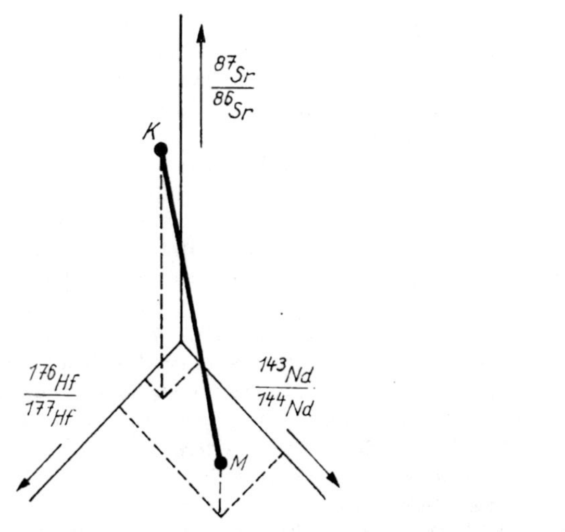

Abb. 4. Räumliches $^{87}Sr/^{86}Sr$-$^{143}Nd/^{144}Nd$-$^{176}Hf/^{177}Hf$-Diagramm und Mischungslinie (MK) von Mantel- und Krustenmaterial

Neodym. Gleichzeitig aber bedingen hohe Ionisierungspotentiale des Hafniumes und Lutetiums erhebliche Schwierigkeiten bei der Massenspektrometrie. Abgesehen von den meßtechnischen Schwierigkeiten ist aber zu erwarten, daß das Lutetium-Hafnium-System Anwendung in der isotopengeochemischen Praxis finden wird. Vermutlich werden mit Hilfe eines räumlichen Diagramms unter Einbeziehung der Hafniumkoordinate (Abb. 4) entscheidende und endgültige Urteile über Mantelherkunft in bisher unklaren und strittigen Fällen möglich sein.

Wir hoffen, daß die neuen Methoden der kombinierten Anwendung natürlicher Isotope dem Streben der modernen Geologie zu ihrem eigentlichen Ziel, dem Auffinden von Bodenschätzen, beitragen werden.

Eingegangen am 15. 6. 1984

Literatur

[1] *V. S. Lepin, T. V. Egorova, M. N. Maslovskaja, S. B. Brandt,* Izotopnyj sostav stroncija kak indikator smešenija v geochimičeskich processach. In: Fizičeskie i chimičeskie metody issledovanija gornych porod i mineralov. Irkutsk: VSF SO AN SSSR 1977, S. 77

[2] *M. N. Maslowskaja, W. S. Lepin, S. I. Kostrowizky, T. I. Kolosnizyna, L. W. Pawlowa, B.. M. Wladimirow, S. B. Brandt,* ZfI-Mitt. **29** (1980) 165

[3] *S. B. Brandt, B. M. Vladimirov, M. N. Maslovskaja, V. S. Lepin, S. I. Kostrovickij* u. a., Sov. geol. **1981**, Nr. 1, S. 94

[4] *R. Vogel,* Heterogene Gleichgewichte. Leipzig: Geest u. Portig 1959

[5] *R. A. Exley, A. P. Jones,* Contrib. Mineral. Petrol. **83** (1983) 288

[6] *D. L. Anderson,* Earth Planet. Sci. Lett. **57** (1982) 1

[7] *A. A. Hofman, W. H. White,* Earth Planet. Sci. Lett. **57** (1982) 421

[8] *D. de Paolo, G. J. Wasserburg,* Geophys. Res. Lett. **3** (1976) 743

[9] *B. B. Jahn, J. Bernard-Griffith, R. Charlot,* Earth Planet. Sci. Lett. **48** (1980) 171

[10] *S. B. Jacobsen, G. J. Wasserburg,* Earth Planet. Sci. Lett. **50** (1980) 139

[11] *G. Faure, J. Powell,* Strontium isotope geology. Berlin – Heidelberg – New York: Springer Verlag 1972

[12] *S. B. Brandt, V. S. Lepin, V. N. Solodjankina,* Sov. geol. **1983**, Nr. 5, 112

[13] *R. Vollmer, C. J. Hawkesworth,* Earth Planet. Sci. Lett. **47** (1980) 91

[14] *P. J. Patchett, M. Tatsumoto,* Nature **288** (1980) 571

Isotopenpraxis 21 (1985) 5, S. 169 bis 175

On Some Problems of the Lead Isotope Method and Its Application in Geology

B. G. Amov

(Institute of Nuclear Research and Nuclear Energy Bulgarian Academy of Sciences Sofia)[1])

Some new trends in the mathematical modelling of the lead isotope evolution in connection with the dynamic development of the earth's crust and with the exchange of materials between the crust and the mantle are discussed. The mathematical modelling of the derivatives with respect to time of U/Pb and Th/Pb ratios $\left(\frac{d\mu}{dt} \text{ and } \frac{dW}{dt}\right)$ *instead of* μ *and* W *give the possibility to describe a wider variety in change of* μ *and* W *with less unknown parameters. It is possible from the isotopic ratios of the lead in a sample to determine the two model ages (uranogenic and thorogenic) and a model parameter. The values of this parameter indicate the time of enrichment (ancient or younger) in U/Pb, which is an extremely important index for identifying the lead source-type. The differences in the uranogenic and the thorogenic model ages, due to the different geochemical behaviour and migration abilities of U and Th at metamorphic processes, provide additional information about the genesis of ores and the source-type of the lead.*

Es werden einige neue Trends in der mathematischen Modellierung der Blei-Isotopenauswertung im Zusammenhang mit der dynamischen Entwicklung der Erdkruste und mit dem Stoffaustausch zwischen der Kruste und dem Mantel diskutiert. Die mathematische Modellierung der Ableitungen hinsichtlich der U/Pb- und Th/Pb-Verhältnisse $\left(\frac{d\mu}{dt} \text{ und } \frac{dW}{dt}\right)$ *in Abhängigkeit von der Zeit statt* μ *und* W *ermöglichen die Beschreibung einer größeren Vielfalt von Veränderungen von* μ *und* W *bei einer geringeren Zahl unbekannter Parameter. Auf Grund der Blei-Isotopenverhältnisse in einer Probe ist es möglich, die beiden Modellalter (das urangenetische und das thoriumgenetische) sowie einen Modellparameter zu bestimmen. Die Werte dieses Parameters weisen die Anreicherungszeit (älteren oder jüngeren Datums) an U/Pb aus, ein außerordentlich wichtiger Index für die Identifizierung der Art der Bleiquelle. Die Unterschiede im urangenetischen und thoriumgenetischen Modellalter, die auf dem unterschiedlichen geochemischen Verhalten und Migrationsfähigkeit von U und Th bei den Stoffwandlungsprozessen beruhen, liefern zusätzliche Informationen über die Erzgenese und die Art der Bleiquelle.*

Keywords

age estimation; continental crust; earth crust; isotope dating; isotope ratio; lead isotopes; mathematical models

1. Introduction

The lead isotope method is one of the oldest methods of geochemical investigation based on the variations in isotopic composition of the elements in nature. After the first investigations of *Nier* in 1938–1941 [1–3] the method was developed in two versions: the U–Th–Pb method for absolute age determination and the lead isotope method, itself, providing information not only for the age, but for

[1]) Address: 1184 Sofia, Bulgaria

the genesis of ores and rocks, for the origin of lead from different sources.

A very important priority of the method over other isotope methods, used in geochemistry, is the fact that the change in lead isotopic composition is a result from the decay of three natural radioactive isotopes with different decay constants, belonging to two elements — U and Th. Consequently, the variations in the isotopic composition of the lead contain a more complete information about the geological and geochemical processes in connection with the global and particular fractionation of the chemical elements throughout the earth's history. Due to the different decay constants of the ^{238}U and ^{235}U, it is possible to identify the changes in the U/Pb ratio in time (ancient or younger), which is a characteristic feature of the lead source-typ, for example, continental crust, oceanic crust or mantle. The differences in the variations of uranogenic and thorogenic lead isotopic ratios provide information about the geochemical conditions and geochemical behaviour of U, Th and Pb in the lead source. This information is useful for the determination of origin and genesis of ores and rocks.

A disadvantage of the method is the lower precision of the lead isotope analyses in comparison with some other isotopes analyses. This is due to the difficulties in the mass-spectrometric isotopic analyses of lead due to its physical and chemical properties. Therefore, earlier, when the precision of the isotopic analyses was about 1%, the method was known as a "rough lead isotope method". For a 1% precision the features of the method retain hidden. However, in the past 10—15 years, owing to the progress in mass-spectrometry, and especially in the emitter method for thermal ionization of lead [e.g. 4—6], the precision of the lead isotope analyses was increased to 0·1%. The use of absolute lead isotope standards [7, 8] makes possible the normalization and the comparison of the results from different laboratories, which was impossible earlier. These new developments of the method considerably increased its possibilities for application in the field of geochemistry and geology.

2. Mathematical models of lead isotope evolution

Different mathematical models of lead isotope evolution were developed and used for interpretation of lead isotope data. In some of them constant values of the U/Pb and Th/Pb ratios (e.g. single-stage models) and a discontinuous evolution in two (two-stage models) or more stages (multi-stage models) have been assumed [e.g. 9—11]. In some more recent models attempts have been made to relate the lead isotope evolution to the geodynamic development of the crust and mantle and the interaction between them. The variations in the lead isotopic ratios and a certain regularity in their evolution have been accounted for by mixing and homogenizing of materials from different layers of the crust and the mantle [12—15]. New ideas on the lead isotope evolution in different zones of the crust and in the mantle in connection with their dynamic development and interaction have been proposed in the Plumbotectonics model [16, 17], which is essentially a multi-stage model of discontinuous evolution. However, the big number of unknown parameters in the mathematical equations is a disadvantage of the multi-stage models. Due to the common parameters of the two equations for radioactive decay of ^{238}U and ^{235}U, it is possible to determine only two unknown parameters. For determination of other parameters it is necessary to assume additional conditions, which very often are arbitrary. The equations of the single-stage model contain only two unknown parameters, but now it is well known that this model is very rough.

The models of continuous evolution make possible a description of the main trend in the evolution with less unknown parameters. For this purpose sutable mathematical functions, which express the change in U/Pb and in Th/Pb ratios (present-day values — $\mu = {}^{238}U/{}^{204}Pb$ and $W = {}^{232}Th/{}^{204}Pb$) in time, are assumed. The problems from the point of view of the continuous evolution has been considered in few models. In 1955 *Bate* proposed a model for continuous differentiation of the earth's crust [18]. Later *Sinha* and *Thilton* [19] proposed a model of the continuous and linear change in the U/Pb ratio, which was further developed by *Cumming* and *Richards* [20].

Now it is well known that, as a result of chemical fractionation, U, Th and Pb are accumulated predominantly in the earth's crust so that the μ and W ratios have increased in the course of the geologic time. In the models of discontinuous evolution this trend is described with a two-stage or multi-stage change in μ and W ratios and in the model of continuous evolution — with the linear change [19, 20] (constant rate of change) in μ and W. The acceptances of constant values or constant rate of change of μ and W are very rough approximations to the real processes, but they follow from the requirement to use a minimum number of unknown parameters in the mathematical model equations.

In a dynamic model [21—23] a new fenomenological approach to the mathematical modelling of the continuous lead isotope evolution was proposed. Instead of μ and W ratios their derivatives with respect to time $\left(\frac{d\mu}{dt}\right.$ and $\left.\frac{dW}{dt}\right)$ are modelled with a suitable mathematical function. The advantage of this approach is the possibility to describe a wider variety in change of μ and W with the aid of a function with fuwer unknown parameters (not more than two).

On the basis of some general trends in the lead isotope evolution, related to the dynamic development of the earth's crust and with the exchange of materials between the crust and the mantle [12—16], as a suitable function for the change in $\frac{d\mu}{dt}$ and $\frac{dW}{dt}$ the two-parametric function is accepted

$$\nu = -\frac{d\mu}{dt} = A(t_0 - t)\, e^{-\frac{t_0 - t}{t_0 - t_m}} \tag{1}$$

where: t is the geologic time; t_0, the age of the earth; A, acceleration factor (dimension t^{-2}); t_m, the time when the μ rate of change was at its maximum. At $t > t_m$ the function (1) is increasing and at $t < t_m$ decreasing (Fig. 1).

The choice of this function is based on the following arguments. If we assume that the temperature is the main factor determining the intensity of the chemical fractionation and the exchange of materials between the different crust and mantle layers, then it would be logical to assume that the μ and W rates of change $\left(\frac{d\mu}{dt} \text{ and } \frac{dW}{dt}\right)$ increased in the initial stage and, after attaining a maximum, started decreasing because of the earth surface cooling.

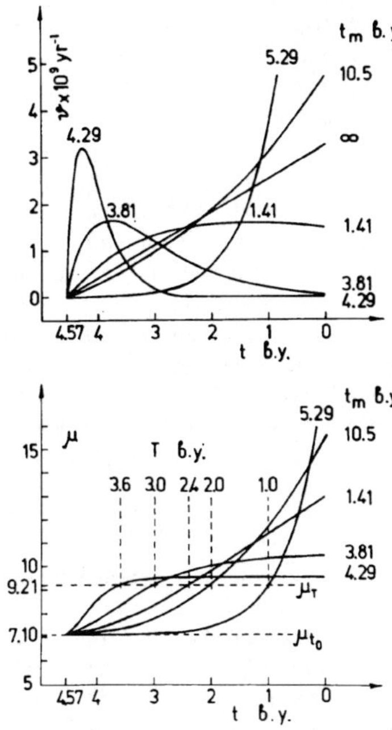

Fig. 1. Changes of ν and μ with geologic time t at different values of the model parameter t_m

The maximum was attained earlier in the upper-crust layers, whereas in the lower ones it was attained later or not at all, since the latter cooled slowler.

Geophisically, the transition from increasing to decreasing μ and W rates of change in a given layer of the crust can be connected with its having been formed from mantle and reworked older crustal materials in accordance with the Plumbotectonics model [16, 17]. The intensity of the chemical fractionation and exchange processes should be maximum near a thermal boundary of high temperature gradient or a phase boundary. After the formation of a layer of the crust the influx of new materials has decreased because of its cooling and hence, the μ and W rates of change have decreased.

Consequently, t_m is a dynamic parameter, the values of which are connected with the age of formation of the crust layers which are the source of lead. Because of the mixing of materials from different layers of the crust and mantle, the values of t_m can be accepted more really as effective (weighted mean) values, that depend on the amount of materials from an older or younger crust or from the mantle.

In this manner the main trend in the lead isotope evolution, related to the dynamic development of the earth's crust and the mantle and with the interaction between them, can be mathematically described as a continuous evolution with less unknown parameters, instead of as a discontinuous multi-stage evolution [16].

The mathematical equations of the model, deduced at these assumptions are more general and include as particular cases, at given values of the model parameters, the equations of some other models. For example, at $t_m = t_0$ they are transformed into the equations of the single-stage model. If we assume that the change of μ started not at the moment t_0, but later, at a moment t_1, at $t_m = t_0$ they are transformed into the equations of the two-stage model. At $\nu = -\frac{d\mu}{dt} = $ const. the change of μ is linear.

3. The model parameters as indicators of the lead origin

The interpretation of the lead isotope data according to the single-stage model showed variations in the μ-value. The lower μ-values are attributed to the lead of mantle origin and the higher – to the lead of crustal origin. In the two-stage model these differences are described with a variation in the μ-value of the second stage and in the model of linear change in μ – with a variation in the linear coefficient of change in μ or in the present-day value of μ. However the μ-values, calculated according to the single-stage equations of the uranogenic lead, are different. Usually the μ-values, calculated according to the equation of the $^{207}Pb/^{204}Pb$ ratio are lower and these, calculated according to the equation of $^{206}Pb/^{204}Pb$ are higher. These differences show that μ has not been constant in the course of the geologic time. On account of the shorter decay period of ^{235}U the value of μ, calculated according to the equation of $^{207}Pb/^{204}Pb$, is an indicator for a lower μ-value in the initial stage of the evolution and those, calculated according to the equation of $^{207}Pb/^{204}Pb$ shows a later increase in μ. This trend is relevant both to the crust and to the mantle, however the lower values of $^{207}Pb/^{204}Pb$ for mantle lead show a later increase in μ in the mantle in comparison to μ in the crust.

The dynamic model [21, 22] describes the lead isotope evolution at different rates of change in μ and W in the course of geologic time. Fig. 1 shows the change in $\nu = -\frac{d\mu}{dt}$ and μ at different values of the model parameter t_m. An early maximum of $\frac{d\mu}{dt}$ and a rapid increase in μ in the initial stage and a subsequent slowing are typical for the older continental crust, whereas for the younger crust and mantle $\frac{d\mu}{dt}$ attained the maximum later or not at all and μ increased slowly initially and faster later. It is seen on this figure that instead of A it is possible to use another model parameter – T, with a clearer physical significance. T is the geologic time when the μ ratio having started from the initial value of μ_{t_0} (e.g. 7·1) and increasing continuously, reached a definite value μ_T (e.g. 9·21). The values of T and t_m are interdependent and are higher for the older continental crust and lower for the younger crust and mantle. The value of T or t_m is therefore a characteristic feature of the lead source.

Fig. 2. shows the diagram $^{207}Pb/^{204}Pb - ^{206}Pb/^{204}Pb$ for Phanerozoic age with the growth curve at $T = $ const. and isochrons at $t = $ const., calculated according to the dynamic model [21, 22] and the growth curve of the Plumbotectonics model [16] for upper crust, orogene and mantle. Some data of modern precision (0·1%) for uranogenic lead isotopic ratios in different types volcanic rocks [24–31], ores [32, 40, 42, 45] and sediments [25, 32], very young in age (Mesozoic-Cenozoic), are ploted on the diagram as well. The lead in these rocks and ores is of different origin. Mantle origin is assume for the oceanic and some continental volcanic rocks [16, 24–31, 33–35], whereas sediments and ores are of predominantly crustal origin [16].

It is seen that the data for oceanic volcanic rocks from the mid-ocean ridge and oceanic islands are located between the growth curves with lower T values ($T = 2{\cdot}6$ to 3·0 b.y.) which is in agreement with their mantle origin. The data for ores and sediments are located between the growth curves with higher T values ($T = 3{\cdot}0 - 3{\cdot}4$ b.y.)

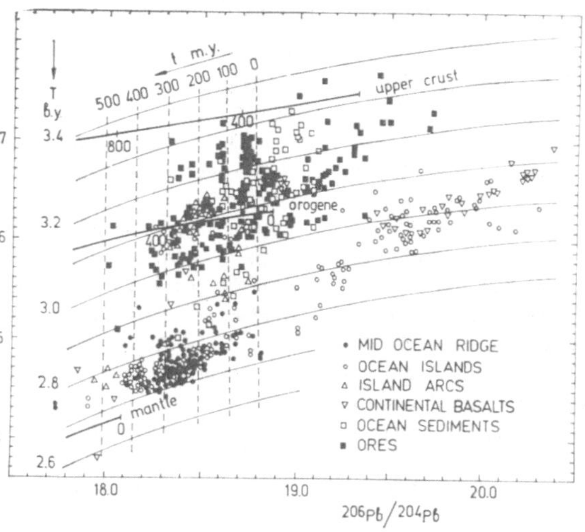

Fig. 2. $^{207}/^{204}$Pb vs. ^{206}Pb/^{204}Pb diagram with growth curves at T = const. and isochrons at t = const. for Phanerozoic age, calculated according to the dynamic model [22]

Some data for the isotopic ratios of uranogenic lead in different types volcanic rocks, ores and sediments are ploted and for comparison the growth curves of the Plumbotectonics model [16] for upper crust, orogene and mantle are given.

which is an indication for their predominantly curstal origin.

On the ground of these isotopic data and in comparison with the growth curves of the Plumbotectonics model [16] it is seen that for the different zones the model parameter T has the following values: $T = 2{\cdot}7$ b.y. for the mantle; $T = 3{\cdot}1$ b.y. for the orogene; $T = 3{\cdot}35$ b.y. for the upper continental crust. It could be assumed that in most cases the intermediate T values (those between the values for the upper crust and for the mantle) reflect different quantitative shares of crustal and mantle materials resulting from repeated mixing between them during the Earth's history. This assumption explained logically the values spread of the ^{207}Pb/^{204}Pb ratio and the model parameter T for ore, sediments and some volcanic rocks.

It is seen from Fig. 2 that recently the lead isotope method is very useful for identifying the source-type of rocks and ores. In comparison with Rb-Sr and Sm-Nd methods, which are also used for this purpose, it is not necessary to construct an isochron to determine the initial isotopic ratios. Because of the small change in the ^{207}Pb/^{204}Pb ratio, owing to the shorter decay constant of the ^{235}U, the conclusion for lead origin may be drawn only from the isotopic data of a single analyzed sample.

4. Some differences in the variations of the uranogenic and thorogenic lead isotopic ratios

Due to the different chemical and geochemical properties of U and Th the changes in U/Pb and in Th/Pb ratios at global processes of chemical fractionation during the Earth's crust development or at metamorphic processes in the separate parts of the Earth are different.

It is known that U occure in nature in two stable valency states — +4 and +6, whereas Th is always in the +4 state. U^{6+} and Th^{4+} are similar in their chemical properties and unlike U^{6+}. Under reductive conditions at great depts U^{4+} predominates, due to which the U and Th migration properties are similar. But under oxidative conditions U valent state changes from + 4 to +6 and it migrates mainly as UO_2^{2+}. The 6-valent U chemical compounds are soluble, in water, hence U^{6+} migrates by means of water solutions. Therefore, under the oxidative conditions in the upper earth's crust U and Th migration abilities differ considerably. The changes in Th/U ratio caused some differences in the evolution and in the isotopic anomalies of the uranogenic and thorogenic lead.

Many experimental data show that Th/U ratio is higher in the crust and lower in the mantle [e.g. 20, 24, 25, 31, 36]. By calibration of the dynamic model [21, 22] it was established, that during the initial stages of the evolution the Th/Pb ratio increased more rapibly than the U/Pb ratio but later the case was reversed. A correlation between the Th/U ratio in time $T(W_T/\mu_T)$ and the values of the model parameter T emerges

$$W_T/\mu_T = 1{\cdot}969 + 0{\cdot}639 \times T(\text{b.y.}) \tag{2}$$

On the basis of this relation a new diagram for the thorogenic lead ^{208}Pb/^{204}Pb versus T (Fig. 3) was proposed [22]. It is seen that the slope of the isochrons on the diagrams of the uranogenic (Fig. 2) and the thorogenic (Fig. 3) lead is different, that reflect the differences in the evolution of the uranogenic and the thorogenic lead. The evolution of the thorogenic lead is better expressed by using the diagram ^{208}Pb/^{204}Pb vs. T that the commonly used ^{208}Pb/^{204}Pb vs. ^{206}Pb/^{204}Pb diagram, since the variations in ^{208}Pb/^{204}Pb and in ^{206}Pb/^{204}Pb very often have different trends, due to the different geochemical behaviours of U and Th at metamorphic events.

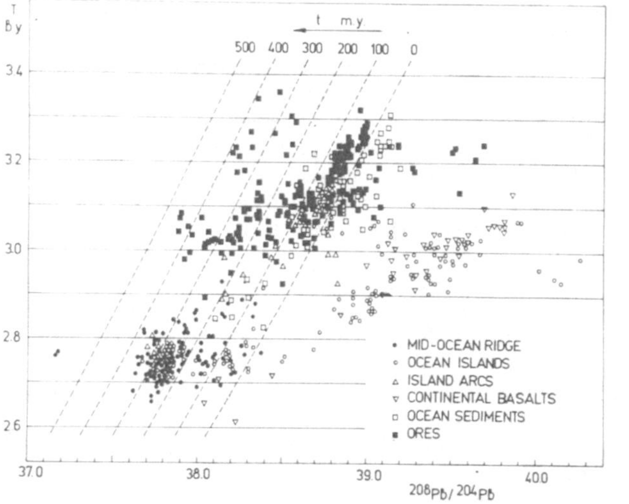

Fig. 3. ^{208}Pb/^{204}Pb vs. T adiagram with growth curves at T = const. and isochrons at t = const. for Phanerozoic age, calculated according to the dynamic model [22]

Some data for the isotopic ratio of thorogenic lead in different types volcanic rocks, ores and sediments are ploted.

It is seen, that on Fig. 2 and Fig. 3 the lead isotope data for oceanic volcanic rocks are located on the both side of the zero-isochrons. An interesting fact is that all data for mid-ocean ridge basalts (MORB) are located on the left side of the zero-isochrons, which indicates a depletion in U/Pb and in Th/Pb of their sources, and most of the data for oceanic island basalts (OIB) are located on the right side of the zero-isochrons which indicates an enrichment in U/Pb and in Th/Pb. This corresponds to trace element studies, showing that these two types of mantle

Kurznachrichten / Newlets

Altersdatierung:
DIAMANTEN MEHR ALS 3 MRD. JAHRE ALT

Ueber das Entstehungsalter von Diamanten herrschen haeufig Kontroversen. Da Diamanten aus reinem Kohlenstoff bestehten, kann man eine Altersdatierung nicht direkt vornehmen. An Granateinschluessen in einigen gefundenen Diamanten konnte der amerikanische Geochemiker S.H. Richardson eine Altersdatierung mit Hilfe der Sm-147/Nd-147- Methode vornehmen, wobei allerdings kristallographisch zunaechst geprueft werden musste, ob auch die Granateinschluesse syngenetisch -also gleichzeitig- mit der Bildung der Diamanten entstanden sind. Die bisher nach der Sm-147/Nd-143-Methode ermittelten Alter der Diamanten liegen zwischen 3,2 ... $3{,}4.10^9$a, haben sich also bereits im fruehen Praecambrium gebildet. (Bild der Wissenschaft (1985) 1, S. 11)

Vulkane:
URANAUSSTOSS BEI VULKANAUSBRUECHEN

Am 18.5.1980 fand eine Eruption des Monat St. Helens statt. Etwa 8000 Mt Staub und Asche wurden in die Atmosphaere gestossen. Nach Fruchter und Mitarbeitern (Science 209(1980) 1116) enthielt das Eruptionsmaterial $1{,}1x10^{17}$ Bq ^{222}Rn, $2.3x10^{15}$Bq ^{40}K, $1{,}2x10^{14}$Bq ^{226}Ra nach $2{,}5x10^{13}$Bq ^{232}Th. Unter der Annahme, dass sich ^{226}Ra mit ^{238}U im radioaktiven Gleichgewicht befand, waren in der vulkanischen Asche auch $1{,}2x10^{14}$Bq ^{238}U. Bei einer homogen Verteilung ueber die Haelfte der Erdoberflaeche, ergab sich ein Fall-out an ^{238}U $4{,}8x10^{-5}$Bq/cm^2. In der Universitaet von Arkansas in Fayetteville, Arkansas (USA) wurden in der Zeit von 1980-1982 Regen- und Schneeproben gesammelt und nach einer Anreicherung deren Urankonzentrationen mit einem α-Spektrometer bestimmt. Nach der Eruption wurde in den Sommermonaten 1980 -die Vulkanaktivitaet hielt an- ein deutlicher Anstieg des ^{238}U-Gehaltes der Niederschlagsprobe beobachtet. Den hoechsten Gehalt an ^{238}U besassen die Proben vom 17. Juli und 2. September 1980. Er entsprach mit 4,4 ... bzw. $4{,}1x10^{-3}$Bq/l etwa dem berechneten Wert von $4.8x10^{-3}$Bq/l. Aus der Bestimmung der Verhaeltnisse ^{234}U/^{238}U bzw. ^{235}U/^{238}U kann auf den Ursprung der Urankontamination geschlossen werden. (nach: P.K. Kuroda u.a. J. Radioanal. Nucl. Chem. - Artic. 84(1984)23)

Activation analysis:
NUCLEAR ACTIVATION METHODS FOR THE CHARACTERIZATION OF FOSSIL FUELS

Due to the anticipated decline in the world reserves of oil and gas, coal and oil shale are expected to be major alternative fuels in the next century. However, both coal and oil shale are "dirty" fuels. It is important to have reliable analytical methods which can monitor the undesirable inorcanic constituents at various stages of production and utilization. R.A. Naelkarei presents a review on the use of neutron activation analysis (NAA) for the analysis of coal, oil shale, tar sands and petroleum. Fast NAA has been widely used for the determination of oxygen, and to a limited extend, of other elements as nitrogen and silicon. Reactor NAA followed by instrumental counting or after radiochemical sperations is discussed. Thermal and epithermal neutrons are both used. Limited use of the Cf-252 sources has been made. On-line bulk coal analyzers has been built for coal blending, combustion control, and coal quality control. Photon-activation analysis can determine over thirty elements, many of them not possible to do by NAA. Round-robin analyses of standard coal, fly ash or oil shale indicate nuclear activation methods are comparable in accuracy and precision to X-ray fluorescence or atomic spectrometric methods for most elements. Nuclear analytical methods appear to be well-suited for multielement and many times nondestrictive analysis of fossil fuels. (. J. Radioanal. Nucl. Chem.Artic. 84(1984)1, pp. 67-87)

Kernenergie:
FRANZOESISCHER ATOMSTROMANTEIL IM NOVEMBER 64.7%

Im November 1984 betrug die Stromproduktion in Frankreich 26,07 Mia kWh. Davon stammten 64,7% aus Kernkraftwerken, 19,5% aus Wasserkraftwerken und 15,8% aus klassisch-thermischen Kraftwerken. (SVA-Bulletin (1985) 1)

30% KERNENERGIE IN BULGARIEN

Das Kernkraftwerk Kosloduj mit seinen 4 aus der Sowjetunion gelieferten 440-MWe-Bloecken hat 1983 annaehernd 30% der gesamten in Bulgarien verbrauchten Elektrizitaet erzeugt. Dadurch steht Bulgarien gegenwaertig an 3. Stelle in der Pro-Kopf-Atomstromerzeugung der Welt. Nach der Inbetriebnahme von zwei weiteren Bloecken, die fuer 1985 und 1986 vorgesehen ist, wird der Kernenergiestrom einen Anteil von ueber 40% erreichen.(Atomwirtschaft/Atomtechnik 30 (1985) 1, K+U II)

Isotopenproduktion:
NEUE ISOTOPENFABRIK

Im Kernforschungszentrum Karlsruhe wurde eine neue Anlage zur Erzeugung radioaktiver Isotope fuer die medizinische Diagnostik in Betrieb genommen. Durch dieses sogenannte Kompaktzyklotron konnte die Produktionskapazitaet der BRD fuer medizinische Radioisotope wesentlich erweitert werden. In der Nuklearmedizin werden komplizierte Stoffwechselvorgaenge mit Hilfe der Gammastrahlung gezielt verabreichter radioaktiver Praeparate untersucht. Eine wesentliche Voraussetzung fuer diese Untersuchung ist die Kurzlebigkeit der verwendeten Radionuklide. (Atomwirtschaft/Atomtechnik 29 (1984) 12, K+U III)

Endlagerung:
FELSLABOR GRIMSEL

Das Felslabor Grimsel in der Schweiz, in dem nach einer zwischen der BRD und der Schweiz getroffenen Vereinbarung, gemeinsame Untersuchungen zur Eignung von Granit als Endlagerformation fuer radioaktive Abfaelle durchgefuehrt werden sollen, wurde termingerecht fertiggestellt und Wissenschaftlern uebergeben. In knapp einem Jahr wurde das etwa 800 m lange Labor in den Granitfelsen des Grimsel-Massivs gefraest. Die Beteiligung der BRD am Grimsel-Projekt dient der Abrundung des Forschungs- und Entwicklungsprogramms zur Endlagerung. Nach wie vor wird in der BRD die Loesung verfolgt, hochradioaktive Abfaelle in tiefen geologischen Salzformationen einzulagern. (Atomwirtschaft/Atomtechnik 29(1984)12, K+U IV)

Energiewirtschaft:
IN BAYERN 60% KERNENERGIE

Nach einer Meldung wurden im Juli 1984 knapp 60% des bayerischen Stromverbrauchs aus Kernenergie gedeckt. Die Vorteile dieser ueberdurchschnittlichen Kernenergienutzung des Landes Bayern werden am Beispiel des Umweltschutzes und der Strompreise deutlich: Eine Reduzierung der SO_2-Emissionen aus den bayrischen Kraftwerken von 1976 bis 1983 um mehr als 75% hätte ohne die Kernenergie nicht in diesem Umfang erfolgen koennen. Die Kernkraftwerke hatten nicht nur den Strom zu ersetzen, der durch einen Mindereinsatz der konventionellen Kraftwerke und durch Stillegung aelterer Kraftwerksbloecke fehlte, sondern auch den zusaetzlichen Strombedarf zu decken, der sich durch eine Verbrauchssteigerung in diesem Zeitraum um 30% ergab. Der hohe Kernenergieanteil hat auch die bayerische Strompreissituation massgeblich positiv beeinflusst.
(Atomwirtschaft/Atomtechnik 29(1984)12, K+U IV)

Spurenanalytik an Gesteinen:
SILIKATANALYSE MIT EDS-XFA

Geologische Probleme sind in der Regel Vielkomponentensysteme und bestehen aus 6 ... 12 Haupt- und Nebenkomponenten meist niedriger Ordnungszahl. EDS-X7-Analysen an Silikatstandards zeigen, dass unverduennte Pulverpresslinge zwar zaehlstatistisch brauchbare Signale liefern, dass Absorptions- und Fluoreszenzkorrekturen notwendig sind, die aber im Falle der verwendeten Rasberry-Heinrich-Routine nicht zufriedenstellen. Schmelzaufschluesse mit $Li_2B_4O_7$ hoher Verduennung (> 10:1) ergeben ueber weite Konzentrationsbereiche lineare Eichfunktionen, hingegen sich die erreichbare Impulsausbeute bei den Elementen mit Z $\leq$ 12 gering und auch bei langen Messzeiten von ueber 200 sec und hohen Konzentrationen analytisch unbrauchbar. Unter Routinebedingungen ist eine semiquantitative bis quantitative Simultananalyse der Hauptkomponenten (Z≥12) moeglich, der Zeitaufwand fuer Praeparation, Analyse und Auswertung liegt fuer 10 ... 12 Analysenelementen bei 1 sec ... 20 min. Bei Verwendung von Pulverpresslingen ist eine Spurenanalyse der Elemente mit Z≥15 im Konzentrationsbereich von ungef. 20 ... 2000 ppm moeglich, wenn geeignete Primaerfilter und Anregungsbedingungen gewaehlt werden. In jedem Fall - Hauptkomponenten - oder Spurenanalysen - ist eine perfekte Praeparationsroutine und eine genuegend grosse Zahl guter Standardproben unerlaesslich.
W.B. Stern; Zur Simultananalyse von Silikaten (Hauptkomponenten, Spuren) mittels energiedispersiver Roentgenfluoreszenz-Spektrometrie (EDS-XFA). [Fresenius' Z. Analyt. Chemie 320 (1985)1]

Ankündigungen / Announcements

N e u e r s c h e i n u n g e n :

IN-BEAM NUCLEAR SPECTROSCOPY

The volumes contain the materials of invited talks, oral and poster contributions, altogether more than 80 puplications submitted to the symposium from different research institutes of about 20 countries.
The main topic of the symposium was the structure of medium heavy nuclei (spherical, transitional and deformed nuclei; even-even, odd and odd-odd isotopes; single-particle, cluster and collective excitations, intruder states; low-spin and high-spin levels; nuclear shapes; symmetry properties of nuclei; nuclei far from stability etc.)
Vol. I, 396 pp. 240 fig. 36 tab., in English
Vol.II, 424 pp.,176 fig.,38 tab., in English
Akademiai Kiado, P.O. Box 24, H-1363 Budapest

Veröffentlichungen / Publications

Berichte aus dem J.W. Kurtschatow-Institut fuer Atomenergie, Moskau

IAE-4072/3

Oekonomie der Sicherheit der Kernenergetik, 1984, W.A. Zegassow u.a.

IAE-4063/15

Automatisierungskomplex fuer Myonen-Untersuchungen, 1984, N.A. Bessonowa u.a.

IAE-4068/3

Analyse methodologischer Aspekte der Risiko-Untersuchungen, 1984, W.K. Suchorntschkin u.a.

IAE-4073/9

Schichtanalysen der strukturellen Vollkommenheit von monokristallinen Oberflaechenschichten mit Hilfe der asymptotischen Bragg-Difraktion, 1985, A.M. Afanasjew u.a.

IAE-4086/11

Der Einfluss des Wasserstoffs und mikrostrukturelle Besonderheiten auf die physikalischen Eigenschaften von nichtrostenden Staehlen der austenitischen Klasse, 1985, P.A. Platenow u.a.

ZENTRALINSTITUT FUER KERNFORSCHUNG ROSSENDORF

ZfK-527

Adsorption von Metallen auf metallischen Oberflaechen und Moeglichkeiten ihrer Nutzung in der Kernchemie. Ermittlung von Adsorptionsenthalpien mit dem Rechenprogramm AMO - H. Rossbach u.a., 1984, 20 S.

KERNFORSCHUNGSZENTRUM KARLSRUHE

KfK-3036

Kinetische Untersuchungen zur Hydrolyse von Tributylphosphat im PUREX-System - 1984, 53 S., R. Becker u.a.

KfK-3796

Abscheidung von elementarem J-131 an Aktivkohle unter extremen Bedingungen - 1984, 33 S., H. Deuber

KfK-3776

Abscheidung von elementarem J-131 an Aktivkohlen unter Stoerfallbedingungen - 1984, 73 S., H. Deuber

KfK-3793

Berechnung der Wert- und Entropieerzeugung in der Trennduese auf der Basis von Molekularsondenmessungen - 1984, 125 S., W. Ehrfeld u.a.

KfK-3784

Schnelle radiochemische Analysenmethode zur Bestimmung von Tc-99 in Umwelt, Abwasser und Abfluss - 1984, 54 S., H.J. Luxenburger u.a.

KfK-NACHRICHTEN 16(1984)4

Forschungs- und Entwicklungsprojekte im Rahmen des Technologietransfer-Programmes des Kernforschungszentrums Karlsruhe - J. Wuest

Das Carix-Verfahren zur umweltfreundlichen Kationen- und Anionenentfernung aus Trinkwasser - W. Hoell u.a.

Elektrolytische Verfahren auf dem Weg zur grosstechnischen Anwendung - H. Goldacker u.a.

Das Karlsruher Kompaktzyklotron - ein Beschleuniger im Dienste der Medizin und Technik - H. Schweickert

Die Produktlinie SPECTRAN zur hochempfindlichen Messung von Gas- und Fluessigkeitskomponenten in der Umwelt- und Prozesstechnik - H. Breton u.a.

Neuere Entwicklungen auf dem Gebiet der Strahlenschutzmesstechnik - E. Piesch

PACOS - ein preiswertes und flexibles System fuer die Bildverarbeitung - D. Vollath u.a.

ISOCOMMERZ-MITTEILUNGEN (1984)4

Isocommerz, 1115 Berlin, Lindenberger Weg 70

N E U:

Gitterpraeparat fuer die Kontrolle von Gamma-Kamera und Scanner - Titel 26

Markierungsschnur fuer die Anwendung in der Nuklearmedizin - Titel 27

Die Schilddruese - aktuelle funktionsmorphologische Diagnostik, Therapie-Schemen,

Isocommerz-Schriftenreihe Band 14, 1984, von H. Deckert

Ausgewählte Beiträge / Recently published papers

IZOTOPTECHNIKA 27(1984)4 (in Hungarian)
Cholescintigraphy with High Hepatotropic ^{99m}Tc-p-butyl-IDA - I. Szilvási et al.

The Significance of Cholescintigraphy in the Diagnosis of Intrahepatic Space-occupying Processes - I. Szilvási et al.

Application of k_0-standardization Method in Neutron Activation Analysis of Agricultural Samples - L. Andraes et al.

Plant Metabolism of Quaternary Benzo-triazolium Derivatives - G. Klivényi et al.

Ethical Problems in the Medical Use of Radioactive Isotopes - B.L. Sztanyik et al.

Metrological Supervision of Radionuclide Calibrators Used in Nuclear Medical Practice, with Radioactive Standards Samples of ^{99m}Tc and ^{57}Co - A. Szoerényi et al.

ATOMNAJA ENERGIJA 58(1985)4 (all in Russian)
On the I-129 Content of VVER Coolant - L.I. Gedeonov et al.

Development of a Model for the Process of Radionuclide Accumulation of Corrosion Products on Equipment of Boiling Reactor APS's - S.A. Tevlin

Some Problems on the Solidification of Liquid Wastes of Medium-Level A tivity Using Inorganic Binders - A.S. Polyakov et al.

Single Crystal Lithium-Fluoride Detectors - A.I. Nepomnyashchikh et al.

The Application of Polarized Radiation for Improving the Sensitivity of Multielement X-Ray Fluorescence Analysis - A.A. Ter-Saakov et al.

The Kerma Constant and Kerma Equivalent of a Radionuclide Source - V.P. Mashkovich

Sensitivity Analysis of a Model of Surface Food Chains - V. Hanusik et al.

Development of a Method of Processing Radioactive Wastes Including Selective Complexing and Ultrafiltration - V.A. Kichik et al.

Yields of Pu-242 Photofission Pruducts - Vo Dak Bang et al.

Determination of the Sensitivity of Thermoluminescent Detectors to Thermal Neutrons - Yu.N. Zhilov et al.

Neutron Radiation Capture Cross-Sections of Fe-58 for E_n-0,5-2,0 MeV - Yu. N. Trofimov

Total Neutron Cross-Section of Ru-101 and R-102 in the 0,01-1750 eV Neutron Energy Range - V.A. Anufriyev et al.

Integral Benchmark on Lead-Slab Attenuation of 14-MeV Neutrons - S.Yu. Antonov et al.

Detectors for X-Ray Radiaction Spectrometry Prepared from High-Purity Germanium Obtained by a Hydride Method - G.G. Devyatykh et al.

Effect of Neutron-Field Parameters on the Yield of Cf-252 - A.V. Klinov

DOKL. AKAD. NAUK SSSR 280(1985)1 (in Russian)
On the problem of ionizing radiation energy delocalization in a molecular medium - J.G. Kaplan et al.

High resolution Si-29 nuclei NMR in layered silicates and aluminuosilicates - V.J. Proshko et al.

Isotopic composition of hydrogen from minerals and rocks of the Tsagan-Suburga deposit (South Mongolia) - A.P. Berzina et al.

DOKL. AKAD. NAUK SSSR 280(1985)2 (in Russian)
Isotopic zoning of the Kuril island arc according the Nd-143/Nd-144 and Sr-87/Sr-86 relations - D.Z. Zhuravlev et al.

DOKL. AKAD. NAUK SSSR 280(1985)3 (in Russian)
A new method of mass transfer calculation in hetero-phase multicomponent systems - L.P. Kholpanov et al.

The age of Precambrian granitoids in the central part of Cisbaikalia according to the data of Pb-Pb thermoisochronic and K-Ar methods - S.P. Korikovskii et al.

On the isotopic composition of C in CH_4 and its homologs in gases from some coal beds and enclosing rocks of the Donets Basin central region - G.J. Voitov et al.

DOKL. AKAD. NAUK SSSR 280(1985)4 (in Russian)
Investigation of heterophase H/D-exchange in the Zn Ni-H-System - A.S. Lunin et al.

DOKL. AKAD. NAUK SSSR 280(1985)5 (in Russian)
Pecularities of the isotopic composition of CH_4 from oil - and gas - and gas-condensate deposits of Azarbaijan - L.M. Zorkin et al.

CHEMIE FUER LABOR UND BETRIEB (CLB) 36(1985)1
Die wellendispersive Roentgenfluoreszenzanalyse - B. Ritschel

J. PHYSIC. CHEM. 89(1985)1
Intracrystalline site preference of hydrogen isotopes in borax - T.M. Pradhananga et al.

NUKLEON (1984) 4 (all in Czech)
Ustav Jaderneho Vyzkumu/Rez

Autonomous Camac Crate - V. Mičkal et al.

Use of inorganic sorbents in nuclear power present state and prospects - P. Franta et al.

Purification of laundry wastes from nuclear power plants - P. Vaňura et al.

Remote control of Zr corrosion rate - V. Zajic et al.

ATOMKERNENERGIE-KERNTECHNIK 46(1985)1
Verlaengerung von SWR-Brennstoffzyklen - U. Schmidt u.a.

Ueber die Brennstoffnutzung in Salzschmelzen-Reaktoren mit optimierter natuerlicher Konvektion - V.M. Novikov u.a.

Modellmaessige Berechnung der Ausbeute bei der Erzeugung von Myonen - Y.V. Petrov u.a.

Zur Loesung der nichtlinearen Fokker-Planck-Gleichung mit Benutzung orthogonaler Polynome - S. Sgouridis

Einsatzmoeglichkeiten von Waermerohren in der Kerntechnik - J. Unk u.a.

Untersuchungen zum Aufloesungsverhalten von bestrahltem niedrig angereichertem HTR-Brennstoff - G. Reitsamer

Kritikalitaet von Curium-Anordnungen - S. Sahin u.a.

Untersuchung der Neutronenempfindlichkeit und des Alpha-Untergrunds von Uran-233-Spaltkammern zur Messung kleiner Neutronenflussdichten - S.M. Lee

J. RADIOANAL. NUCL. CHEM. - LETT. 93(1985)1

A catalytic activity model based on the structural defects of catalysts I. Co-60 gamma radiation field effect on 1-butene isomerisation on an alumina catalyst - S. Illie et al.

A catalytic activity model based on the structural defects of catalysts II. Alumina catalyst characterisation by radioluminescence method - S. Ilie et al.

Solvent extraction of cadmium (II) with 2-mercaptobenzothiazole into methyl iso-butyl ketone - R.K. Itawi et al.

Determination of heavy metals in welder's working environment using nuclear analytical methods - L. Olah et al.

Chemical states of tritium in solid lithium compounds irradiated with neutrons - H. Kudo et al.

J. RADIOANAL. NUCL. CHEM. - LETT. 93(1985)2

Extraction and separation of mercury from toxic elements arsenic, cadmium, antimony and thallium, and other metal ions - S.M. Hasany

Determination of Cu, Zn and Mn by neutron activation analysis in the chromatin of lymphocytes from patients with skin diseases - I. Horkay et al.

Rapid assay of fluorine in polyethylene by NAA - H. Sorek

Depth distribution of Ce-137, Sr-90 and Pb-210 in sediments of lake Mondsee, Austria - K. Irlweck

J. RADIOANAL. NUCL. CHEM. - LETT. 93(1985)3

A new method for the extraction of Au(III) with ethyl thioacetoacetate into chloroform - S.Z. Khan et al.

Determination of the plutonium concentration by isotope dilution mass spectrometry using Pu-239 as a spike - S.K. Aggarwal et al.

A study on solvent extraction of Cu(II) with 2-mercaptobenzothiazole into chloroform - M. Rajadhyaksha et al.

The presence of antimony in various dental filling materials - A. Molokhia et al.

J. RADIOANAL. NUCL. CHEM. - LETT. 93(1985)4

Separation of elements by means of fractional sublimation of their β-diketonates in the vapours of β-diketones - S.S. Travnikov et al.

Thermochromatographic separation of americium and neptunium hexafluoroacetylacetonates in the flow of a carrier gas saturated with hexafluoroacetylacetone vapour - S.S. Travnikov et al.

On processing standards in radiochemical neutron activation analysis - A.R. Byrne

ATOMWIRTSCHAFT/ATOMTECHNIK 29(1984)12

Siedewassereaktoren in der BRD - W. Keller

Das Kernkraftwerk Gundremmingen B und C - H. Haussmann u.a.

Erfahrungen bei den SWR-Anlagen Kruemmel und Grundremmingen II B - R. Ettemeyer u. a.

Betriebserfahrungen mit Siedewasserreaktoren-Kernkraftwerken - A.J. Huettl

Entwicklungsarbeiten fuer Siedewasser-Reaktoren - P. Kilian u.a.

Die Entwicklung von Brennelementen fuer Siedewasserreaktoren - R. Holzer u.a.

Tendenzen in der Abgabe radioaktiver Stoffe mit Fortluft und Abwasser - M. Skafi u.a.

Tagungsbericht:
Energie in der Geschichte (Bericht ueber das ICOHTEC-Symposium in Bergisch Gladbach)

Statistik:
Betriebsergebnisse der Kernkraftwerkein der EG 1983

Vergleich der Stromerzeugungskosten aus Kernenergie, Kohle, Oel in den USA 1983

ATOMWIRTSCHAFT/ATOMTECHNIK 30(1985)1

Operating Experience with Reprocessing Plants - M. Delange

Ergebnisse, Stand und Perspektiven der Reaktorsicherheitsforschung - H.G. Seipel

Nutzung der Spaltprodukte nach Aufarbeitung ausgedienter Brennelemente - E. Gelfort

Leitlinien zur Beurteilung der Auslegung von Kernkraftwerken - J.B. Fechner u.a.

Decontamination of Radioactive Solutions Using Ultra-Filtration - H. Loewenschuss u.a.

Ein alternatives Konzept zur Zuendung von Fusionsreaktionen - W. Trommsdorff

Tagungsbericht:
Fusionstechnologie (Bericht ueber das "13th Symposium on Fusion Technology (SOFT)" in Varese

ACTA PHYSICA POLONICA A66 (1984)5

Spectroscopy of triatomic hydrogen (H and D) - H. Figger

J. RADIOANAL. NUCL. CHEM. ART. 88(1985)1

Significance of specific activity and a possible universal unit for its definition - K. Svoboda

Sensitivities and interferences in activation analysis with cyclotron-produced fast neutrons - M. Esprit et al.

Simultaneous determination of nitrogen and carbon in silicon by photon activation - M. Fedoroff et al.

Radioanalytical methods in the hot atom chemistry of sulfur in inorganic systems - D.S. Todorovsky et al.

Laboratory of nuclear analysis and control methods with a special purpose reactor - S.M. Neretin

INAA results for metal traces in human whole blood. Frequency and age characteristies of the distribution - L.M. Mosulishvili et al.

Estimation of the effect of thermal and resonance neutron shielding in activation analysis - Yu. V. Yakovlev et al.

Determination of nitrogen by neutron activation in meteorites and rocks - A.K. Lavrukhina et al.

Neutron activation method in the analysis of geological materials for noble metals - E.E. Rakovskii

Substoichiometric analysis in the USSR - G.N. Bilimovich

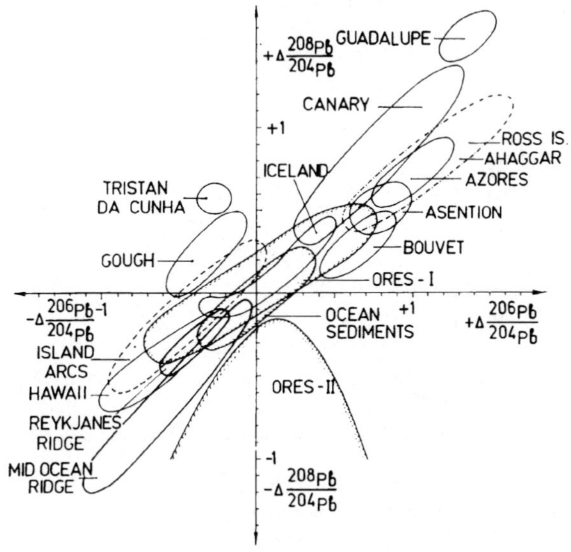

Fig. 4. Fields of deviations of uranogenic and thorogenic lead isotopic ratios from their theoretical values, calculated according to the dynamic model [22], for lead from different types of sources

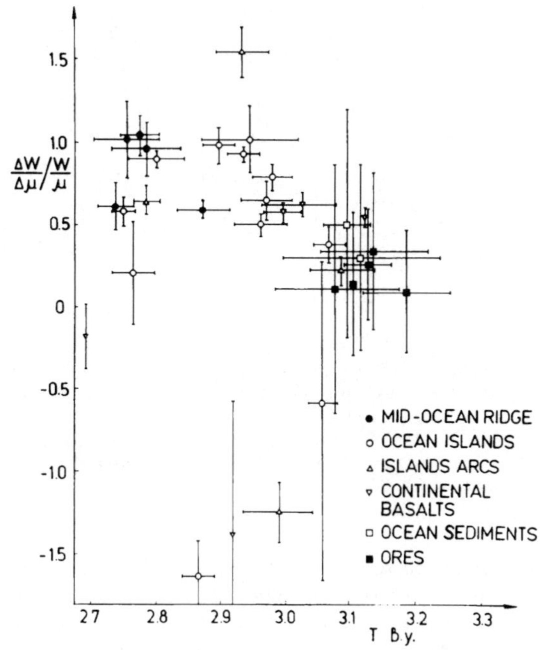

Fig. 5. Relation between the relative change in Th/Pb ratio $\left(\frac{\Delta W}{\Delta\mu} \middle/ \frac{W}{\mu}\right)$ and the model parameter T, calculated according to the dynamic model [22] for lead from different types of sources

reservoirs are depleted and enriched, respectively, in the large-ion lithophile elements [25, 37–39] where U, Th and Pb also belong. Consequently, on the ground of this fact it may be considered that the positions of the zero-isochrons for uranohenic and thorogenic lead also reflect satisfactorily the evolution of the mantle lead.

Fig. 4 shows the fields boundaries where there are deviations in the uranogenic and in the thorogenic lead isotopic ratios from their model values for different types of young ores, sediments and volcanic rocks. For the mantle lead a unidirectional change in uranogenic and thorogenic lead isotopic ratios is typical. These changes are both negative for MORB samples and both positive for most of the OIB samples. But, for some samples from islands (Tristan de Cunha and Gough) and island arcs the deviations in uranogenic ratio are negative but these in thorogenic ratio are positive. The deviations for some ore samples (ORES-I) and for sediment samples are smaller and in the both sides (negative and positive) but the changes in the uranogenic lead isotopic ratio are higher than those in the thorogenic lead isotopic rtio. However, for other types of ores (ORES-II) the deviations in $^{208}Pb/^{244}Pb$ are only negative but those in $^{206}Pb/^{244}Pb$ are on both sides. This could be accounted for with a selective extraction of uranogenic and thorogenic radiogenic lead from the source, for example, under the conditions of a low-temperature hydrothermal activitiy, where the soluble U^{6+} compounds and the radiogenic uranogenic lead could be easilly extracted than the less soluble Th^{4+} compounds and the radiogenic thorogenic lead, respectively.

Fig. 5 shows a rough correlation between the values of the model parameter T and the relative changes in the Th/U ratio $\left(\frac{\Delta W}{\Delta\mu} \middle/ \frac{W}{\mu}\right)$ calculated according to the equations of the dynamic model [22] and the equations of the single-stage model. The lowest T values and the highest $\frac{\Delta W}{\Delta\mu} \Big/ \frac{W}{\mu} \approx 1$ are relevant to the MORB. They are close to 1 which is an indication of the similarity in the geochemical behaviour of U and Th under the reductive conditions in the mantle. The highest T values and the lowest $\frac{\Delta W}{\Delta\mu} \Big/ \frac{W}{\mu}$ (close to 0) are relevant for the ores and sediments, which is another indication of their crustal origin. This feature of crustal-type sources could be related to the oxidation conditions near the earth's surface where a part of the U is in 6+ state, forming compounds easily soluble in water, whereas the compounds of Th are less soluble and therefore the W changes are smaller than the μ changes. The T values and the relative changes in Th/U ratio $\left(\frac{\Delta W}{\Delta\mu} \middle/ \frac{W}{\mu}\right)$ for most of the OIB samples are intermediate between these for MORB and for the samples of ores and sediments. The same trend is observed for some samples of continental alkali basalts too. In some cases this trend could be explained with the assumption of crustal admixtures in the mantle [12, 13, 16, 34, 35] or with crustal contamination of the magmas. But in other cases Nd and Sr isotopes do not confirm these assumption, and to explain the "lead paradox", a core-mantle interaction is assumed [25, 30, 33]. However, it is difficult to explain the different trend in the change of U/Pb and Th/Pb ratios in OIB only with the core-mantle interaction. It seems that the real processes are more complicated.

5. Uranogenic and thorogenic model ages

The mathematical equations of a model describe only a general trend in the lead isotope evolution. This trend depends strongly on the model equations and their calibration to the lead isotope data of samples with known geologic age.

Fig. 6 shows isochrons for the uranogenic lead, calculated according to some recently used models: the two-stage model of *Stacey* and *Kramers* [10] (S-K), the model of linear change in μ and W of *Cumming* and *Richards* [20] (C-R),

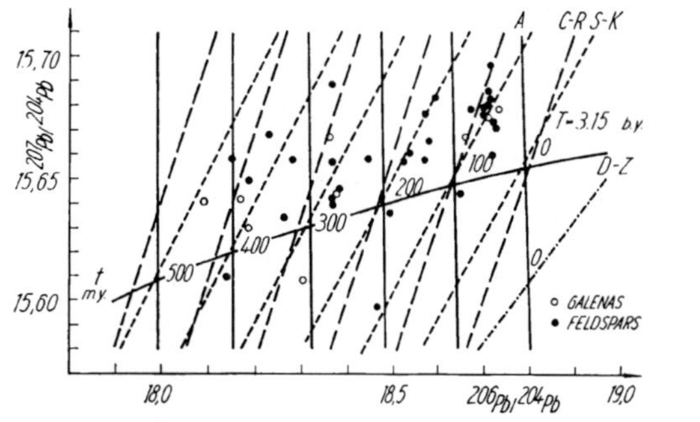

Fig. 6. $^{207}Pb/^{204}Pb$ vs. $^{206}Pb/^{204}Pb$ diagram with isochrons, calculated according to the models of *Stacey* and *Kramers* [10] (S–K), *Cumming* and *Richards* [20] (C–R), *Doe* and *Zartman* [16] (D–Z) and *Amov* [22] (A)

A growth curve at $T = 3{\cdot}15$ b.y. and some data for the isotopic ratios of uranogenic lead in galenas and feldspars from ores and rocks in Bulgaria are ploted.

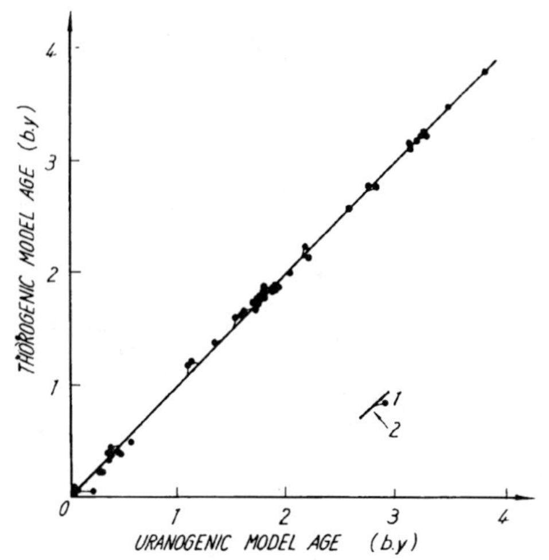

Fig. 7. Relationships between the uranogenic and the thorogenic model ages (1) and the geologic ages (2) of 52 ore deposits in various continents (according to [22])

the Plumbotectonics model of *Doe* and *Zartman* [16] (D-Z) and the dynamic model of *Amov* [21, 22] (A). It is seen that the isochrons of the models (S-K), (C-R) and (A) intersect one another near the growth curve at $T = 3{\cdot}15$ b.y., which is close to the "orogene" growth curve of the Plumbotectonics model [16]. Therefore, the uranogenic model ages, calculated according to these three models, are in these cases close. They differ however when the values of $^{207}Pb/^{204}Pb$ and T are higher (upper crust sources) or lower (lower crust or mantle sources). The reason for these differences are the more simple mathematical equations of the models (S-K) and (C-R) and its calibration by the lead isotope data predominantly of "orogene" sources. The steeper slope of the isochrons of the dynamic model (A) follows from the more general model equations and its calibration to more lead isotope data of different type of sources [21, 22]. The zero-isochron of the model (D-Z) deviate considerably from these of the other models, which shows that the calibration of this model is not appropriate for the purpose of geochronology.

The differences in the thorogenic model ages, calculated according to the different models, are greater, since in the models (S-K) and (C-R) the single growth curves for the thorogenic lead are used. In the dynamic model (A) a dependance of the $^{208}Pb/^{204}Pb$ ratio on the type of the lead sources by means of the model parameter T is used. The calculated thorogenic model age, using the new proposed diagram $^{208}Pb/^{204}Pb$ vs. T (Fig. 3) is closer to the rela geological age in comparison to that, calculated according to the other models (S-K) and (C-R). In the case of using the diagram $^{208}Pb/^{204}Pb$ vs. $^{207}Pb/^{204}Pb$ the isochrons of the models (A) and (D-Z) are close.

Fig. 7 shows a good agreement between the geologic age, ranging from 0 to 3·8, b.y. and the model ages (uranogenic and thorogenic) for most of the 52 ore deposits, whose Pb isotopic data were used for calibration of the model equations [21, 22]. However, the different geochemical behaviours and migrations abilities of U and Th at metamorphic processes cause different change in the isotopic composition of uranogenic and thorogenic lead and in the uranogenic and thorogenic model ages, respectively.

The comparison between the uranogenic and thorogenic model ages reveals three cases:

1) They are approximately equal. This is the normal case.

2) The uranogenic model age is higher than the thorogenic. This is an indication for greater depletion in U/Pb than in Th/Pb of the Pb source as a result of earlier metamorphic processes. Usually, the type-B anomaly ($t_{mod.} > t_{geol.}$) is characteristic in this case, but more often this anomaly refers mainly to the urganogenic lead, whereas the thorogenic model age is closer to the geologic age. There are data showing that urganogenic lead is less radiogenic in the lower than in the upper continental crust, whereas with thorogenic lead these differences are considerably smaller. In the Plumbotectonics model [16] this is accounted for by the depletion of the lower crust and the enrichment of the upper crust in U as a result of high-rank metamorphism. The Th and Pb contents in the lower continental crust do not appear so drastcially reduced. Therefore, when the source of the lead is the lower continental crust the thorogenic model age is closer to the geologic age but uranogenic model age is higher and indicates an earlier metamorphism.

3) The uranogenic model age is lower than the thorogenic. One of the reason may be a greater enrichment of the lead source in U/Pb than in Th/Pb. Usually the type-J anomaly ($t_{mod.} > t_{geol.}$) is characteristic in this case, but

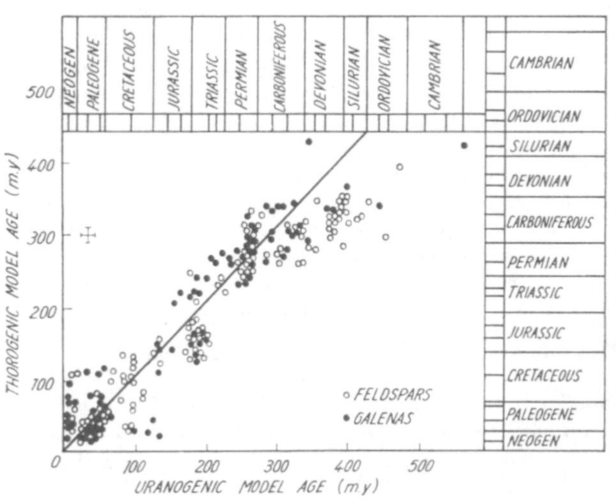

Fig. 8. Relationships between the uranogenic and the thorogenic model ages of lead in feldspars from granitoid rocks and in galenas from ore deposits in Bulgaria

the anomaly for the uranogenic lead is greater too. The source of the lead in this case may be the upper continental crust, usually more enriched in U/Pb than in Th/Pb [16]. But in other cases, especially for some ores (e.g. ORES-II on Fig. 4), the reason for the anomaly may be a selective extraction of uranogenic and thorogenic radiogenic lead from the source under the conditions of low-temperature hydrothermal solutions, where the soluble U^{6+} compounds and the uranogenic radiogenic lead could be extracted more easily than the less soluble Th^{4+} compounds and thorogenic radiogenic lead, respectively [23].

As an example Fig. 8 shows the relationships between the geologic age and the uranogenic and the thorogenic model ages of lead in potash feldspars of granitoids and in galenas of ore deposits in Bulgaria [40—47]. Besides the normal case (uranogenic, thorogenic and geologic ages are equal) there occur the other two cases as well. For the Hercinian granitoids in the Western Balkan Mountains and for the Alpine intrusion in the Central Srednogorie the uranogenic model age is higher than the thorogenic, which indicates origin from the metamorphosed parts of the crust, more depleted in U/Pb than in Th/Pb. The thorogenic model age of the lead in these rocks is closer to their geologic age. But for other type of granitoids in these regions the uranogenic and the thorogenic model ages are equal and coincide with the U-Pb age (Permian). The same is valid for some granitoids and volcanic rocks in Southern Bulgaria, but their model and U-Pb ages are lower. The isopic composition of lead in these rocks is similar to that of lead in ore deposits in the same regions with young Alpinian geologic age. In many cases the lead isotopic data show close genetical relation (common source and age) between igneous rocks and ores. But in other cases the genesis of the ores is quite different. For example, the stratiform ore deposits in the Western Balkan Mountains do not show relationship with any igneous activity, but the lead isotopic data indicate a redeposition of older ore mineralisations. The thorogenic model age is higher than the uranogenic and the geologic ages, which indicates a low-temperature genesis of this type of ores, possibly related with a selective extraction of the uranogenic and the thorogenic radiogenic lead from the source's rocks.

The relatively high values of the $^{207}Pb/^{204}Pb$ ratio (15·60 to 15·70) and of the model parameter T (3·1 — 3·3 b.y.) indicate a predominantly crustal origin of the ore deposits and magmatic rocks in Bulgaria.

These and other examples demonstrate, that the differences in uranogenic and thorogenic model ages and the variations in $^{207}Pb/^{204}Pb$ ratio contain useful information about the origin of the lead from different types of sources, on the genesis of ores, etc.

Received October 4, 1984

References

[1] *A. O. Nier*, J. Amer. Chem. Soc. **60** (1938) 1571
[2] *A. O. Nier*, Phys. Rev. **55** (1939) 153
[3] *A. O. Nier, R. W. Thompson, B. F. Murphey*, Phys. Rev. **60** (1941) 112
[4] *P. A. Akishin, O. T. Nikitin, G. M. Panchenkov*, Geochimia **5** (1957) 425
[5] *B. G. Amov*, Isotopenpraxis **4** (1968) 358
[6] *D. E. Cameron*, Anal. Chem. **41** (1969) 525
[7] *E. J. Catanzaro*, Earth Planet. Sci. Lett. **3** (1967) 343
[8] *E. J. Catanzaro, T. J. Murphy, W. R. Shields, E. L. Garner*, Journ. Res. Nat. Bureau Standards, A. Physics and Chemistry **72**, A3 (1968) 261
[9] *F. G. Houtermans*, Naturwissenschaften **33** (1946) 185
[10] *J. S. Stacey, J. D. Kramers*, Earth Planet. Sci. Lett. **26** (1975) 207
[11] *N. H. Gale, A. E. Mussett*, Rev. Geophys. Space Phys. **11** (1973) 37
[12] *R. L. Armstrong*, Rev. Geophys. **6** (1968) 1975
[13] *R. L. Armstrong, S. M. Hein*, Geochim. Cosmochim. Acta **37** (1973) 1
[14] *R. D. Russell*, Rev. Geophys. Space Phys. **10** (1972) 529
[15] *R. D. Russell, J. D. Birnie*, Phys. Earth Planet. Inter. **8** (1974) 158
[16] *B. R. Roe, R. E. Zartman*, Plumbotectonics. I. The Phanerozoic, in: Geochemistry of Hydrothermal Ore Deposits, *H. L. Barnes*, ed., 2nd ed., New York, N.Y.: John Wiley and Sons. 1979
[17] *J. S. Stacey, B. R. Doe, L. T. Silver, R. E. Zartman*, Plumbotectonics, II A. Precambrian ore deposits, in- Geochronology and Problems of the Ore Formation, pp. 93—106, Moskwa: Nauka 1977
[18] *R. D. Russell, R. M. Farquhar*, Lead Isotopes in Geology. New York, N.Y.: Interscience 1960
[19] *A. K. Sinha, G. R. Tilton*, Geochim. Cosmochim. Acta **37** (1973) 1823
[20] *G. L. Cumming, J. R. Richards*, Earth Planet. Sci. Lett. **28** (1975) 155
[21] *B. G. Amov*, Geol. Balcanica **10**, 3 (1980) 93
[22] *B. G. Amov*, Earth Planet. Sci. Lett. **65** (1983) 61
[23] *B. G. Amov*, Earth Planet. Sci. Lett. **65** (1983) 311
[24] *M. Tatsumoto*, Earth Planet. Sci. Lett. **38** (1978) 63
[25] *S.-S. Sun*, Philos. Trans. T. Soc. London, Ser. **A297** (1980) 409
[26] *S. E. Church, M. Tatsumoto*, Contrib. Mineral. Petrol. **53** (1975) 253
[27] *S.-S. Sun, M. Tatsumoto*, Science **190** (1975) 143
[28] *S.-S. Sun, B. M. Jahn*, Nature **255** (1975) 527
[29] *S.-S. Sun, G. N. Hanson*, Contrib. Mineral. Petrol. **52** (1975) 77
[30] *C. J. Allègre, B. Dupré, B. Lambert, P. Richard*, Earth Planet. Sci. Lett. **52** (1981) 85
[31] *C. J. Allègre, B. Dupré, P. Richard, D. Rousseau*, Earth Planet. Sci. Lett. **57** (1982) 25
[32] *B. R. Doe*, Lead isotope data bank: 2,624 samples and analysis cites, U.S. Geol. Surv. Open File Rep. (1976) 76—201
[33] *C. J. Allègre, O. Brévart, B. Dupré, J.-F. Minster*, Philos. Trans. R. Soc. London, Ser. A **297** (1980) 447
[34] *J. Tarney, D. A. Wood, A. D. Saunders, J. R. Cann, J. Varet*, Philos. Trans. R. Soc. London, Ser. A **297** (1980) 179
[35] *D. L. Anderson*, Earth Planet. Sci. Lett. **57** (1982) 1
[36] *B. R. Doe*, Lead Isotopes. Minerals, Rocks and Inorganic Materials Series. 2. New York: Springer-Verlag 1970
[37] *P. W. Gast*, Geochim. Cosmochim. Acta **32** (1968) 1075
[38] *G. N. Kay, P. W. Gast*, J. Geol. **81** (1973) 653
[39] *J. F. Minster, C. J. Allègre*, Contrib. Mineral. Petrol. **68** (1978) 37
[40] *B. G. Amov, B. Bogdanov, Ts. T. Baldjieva*, Problems of Ore Deposition. 11 (Fourth IAGOD Symposium, Varna, Bulgaria, 1974) p p.13
[41] *B. G. Amov, V. S. Arnaudov, Ts. T. Baldjieva, M. D. Pavlova*, Compt. rend. Acad. bulg. Sci. **29** (1976) 1027
[42] *B. G. Amov, Ts. T. Baldjieva*, Compt. rend. Acad. bulg. Sci. **29** (1976) 1031
[43] *B. G. Amov, Ts. T. Baldjieva, T. Gadjeva, P. Lilov*, Compt. rend. Acad. bulg. Sci. **30** (1977) 1745
[44] *B. G. Amov, J. Mincheva-Stefanova, Ts. T. Baldjieva*, Geochimia, **4** (1979) 535
[45] *B. G. Amov, V. V. Breskovska, Ts. T. Baldjieva, S. S. Evstatieva*, Compt. rend. Acad. bulg. Sci. **32** (1979) 1271
[46] *B. G. Amov, V. Arnaudov, M. Pavlova, P. Dragov, Ts. Baldjieva, S. Evstatieva*, Geol. Balcanica **11**, 2 (1981) 3
[47] *B. G. Amov, V. S. Arnaudov, M. D. Pavlova*, Compt. rend. Acad. bulg. Sci. **35** (1982) 1535

Berichtigung

Heft 7/1984, S. 246, li. Spalte, Zeile 30 von unten: Statt Patienten lies Partien.
Heft 9/1984, S. 363, re. Spalte, letzte Zeile: Statt Kontaminationen lies Kombinationen.

Isotopenpraxis **21** (1985) 5, S. 176 bis 179

Equipment Systems for Radiotracer Studies of Dynamic Phenomena in Technological Processes

R. Malik, J. St. Michalik

(Institute of Atomic Energy)[1]

The electronic measuring system for radiotracer investigations of dynamic phenomena in industrial technological processes and environment protection has been described. The versatile and multi-functional system together with a digital minicomputer is a basic equipment of the specially designed mobily laboratory. The measuring system is currently being provided with the correlation analyses apparatus and the radio remote control system for communication between detectors and the central recording unit.

Es wird ein Apparatursystem für Radiotracermessungen der dynamischen Erscheinungen in industriellen technologischen Prozessen und im Umweltschutz beschrieben. Das vielfunktionelle Meßsystem mit Minicomputer ist Teil der speziellen Ausstattung des Meßwagens, der mit Geräten zur Korrelationsanalyse und mit Funkanschluß zwischen den Detektoren und der Registrierapparatur ausgerüstet ist.

Keywords

CAMAC system; computers; laboratory equipment; multielement analysis; tracer techniques

Introduction

The radioactive tracer method as applied to industrial investigations of dynamics phenomena has been practically used by the Institute of Nuclear Research[2]) since 1960. During the past, over twenty years period, this method has been systematically developed and used in more and more extended range. The Department of Nuclear Methods for Process Engineering, Institute of Nuclear Research (at present a part of Institute of Nuclear Chemistry and Technique), in its many years practice, has used this method, among other ones, for testing the industrial units and installations, investigating technological processes, and carrying out hydrodynamic measurements in tanks and open waters. The tests using radioactive tracers are now applied for practically all phases of investigated medium including fluidized beds, and such tests comprise a wide scope of problems in various industrial branches. In general, to list here exemplarily, there are: metallurgy, chemical industry, power engineering, building material industry, raw materials processing industry, environmental protection, etc.

The concentration of such extended problems in a single department has entailed the need of solving a number of particular problems and, additionally, has brought about a general problem which reduces to the necessity of developing a functionally – versatile measuring system.

In the particular case of the Department of Nuclear Methods for Process Engineering, this versatility means the capability of carrying out the tracer tests on any objects and under any conditions. From the measurements arrangement viewpoint, as has been shown by the former many years practice, such a measuring system should comply with the following requirements:

– adaptable for the testing under industrial and field conditions;
– enabling the recording of single measurement results and, at the same time, suitable for the whole series of programmed tests, i.e. such ones in which, e.g. a change in the technological parameters on the object under test is introduced after the analysis of preceding measurement results;
– enabling the measurements on small objects as well as in cases where the measurement points are spaced one from another by hundreds of metres.

The requirements listed above was settleted by us as the basic design assumptions for the planned measuring system. The project assumed, moreover, that its accomplishment will be implemented gradually with practical checking of the usefulness of constructed system parts at individual stages.

The accomplishment of the developed project to construct the universal industrial-field radiotracer measurements system practically began in 1976. At this time, the Department of Nuclear Methods for Process Engineering, INR, had at its disposal a simple analog equipment con-

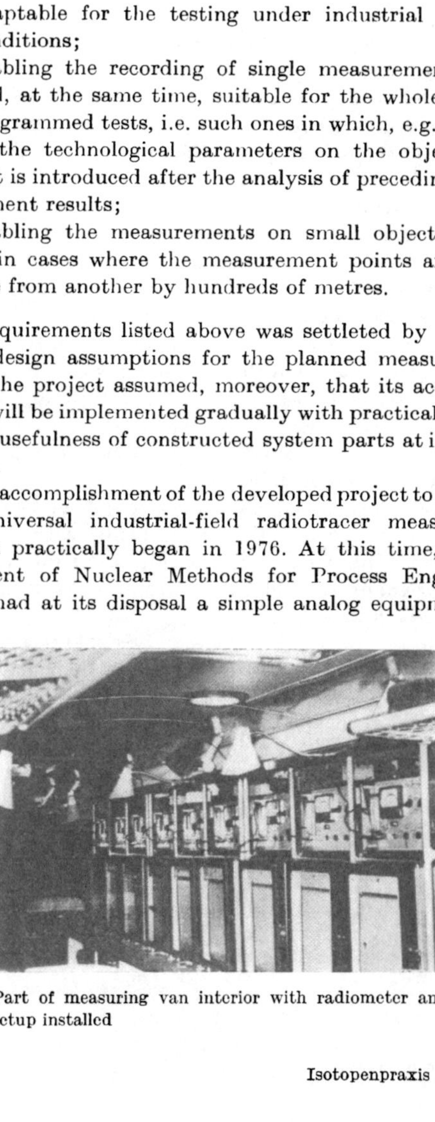

Fig. 1. Part of measuring van interior with radiometer and recorder setup installed

[1]) Address: Świerk, 05-400 Otwock, Poland

[2]) In 1983 the Institute of Nuclear Research replaced by 3 institutes: Institute of Atomic Energy, Institute of Nuclear Chemistry and Technique, Institute of Nuclear Studies.

sisting of 12 independently operating radiometers conected with potentiometric recorders. The set of these radiometers was installed in a van as shown in Fig. 1.

Measuring equipment

The basic problem requiring a general solution was the measuring equipment question. The methodological limitations of analog recording method arose, first of all, from the limited results accuracy. E.g. the mean residence time of a medium in an object under test was burdened by the error:

- 0,8 to 4% in testing the cement kiln process
- 1,0 to 22% in testing the autoclave processes
- 20,0 to 50% in testing the fluidized bed movoment and gas flow.

The analysis of the Department needs has lead to the conclusion that the only acceptable solution is to use digital technique. It was necessary, because of the reasons mentioned above, to accept such temporary solutions that would ensure basic functionality at any stage of the system development and be, partly at least, fragments of the planned system. Such a solution has been provided by the use of home-made (POLON) type "Standard M-ZM-703" medical test sets. These three-channel integrating sets (Fig. 2) still featured drawbacks typical for the analog measurement and recording methods, however, provided the following advantages over the previously used equipment:

- time unambiguity of the recordings of all three channels,
- facility of magnetic recording of pulse frequency and the recording reproduction in laboratory with direct inputting to computer (e.g. into the multi-chanel "Plurimat-Multi 20" system), hence the use of digital technique and the possibility to avoid analog-method-inherent errors.

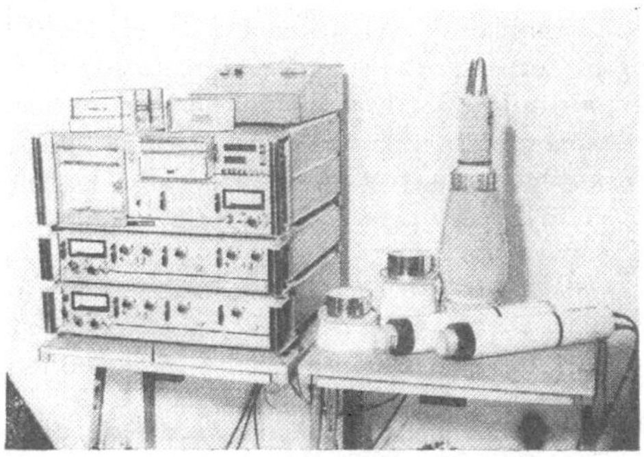

Fig. 2. Type ZM-703 three-channel measuring setup

After several months of use of these sets, design assumptions for digital data recording system have been developed.

In equipment following our assumptions and intended for the operation with just used ZM-703 sets has been constructed by POLON and called IZOLDA, in 1977. The equipment arrangement consisting of five three-fold-ZM-703 and IZOLDA sets made up the first digital measuring set used by the Department for industrial radiotracer measurements. The functionality of this system permits:

- simultaneous digital data recirding from 15 detectors,
- printout and/or punched-tape recording
- independent simultaneous analog recording (integrators and recorders)
- magnetic recording, independent and parallel as well.

The measuring system developed was interesting in its exceptional use of hybrid arrangement operating in parallel both in analog and digital systems (multiscaler).

The digital recording was of sequential type and each sequence contained the information on reading time (from measurement start), detector no. and the number of pulses provided by this detector. These data (in form of punched tape) were, after completion of measurements entered directly to an apropriately programmed computer where the data were processed and the technological parameters of object under test computed. Unlike conventional multichannel analysers operating in a multiscaler mode, the system concerned has an essential novelty, i.e. a programmed interruption time between measurements. This solution enabled the operation in either modes:

a) measurement — measurement — measurement . . .

b) measurement — interruption — measurement . . .

Another feature that differentiates this system from multichannel analysers is practically unlimited number of channels which permits continuous tracking of even so long term processes as e.g. dynamic phenomena in large glass-making units (the experiment duration reaches 300 h).

An essential drawback, from practical point of view, was a relatively long time per channel. The minimum theoretical value of this parameter is 0·1 s. however, it is limited in practice by the operating speed of peripherals and amounts from 1 to 4 seconds per channel (depending on the type of peripheral device and the number of ZM-703 sets connected to the system). This shortcoming prevented the system from being used for investigating the high-speed processes.

In our practice, a number of studies aims at the optimization of technological process conditions, which, provided a proper measurements organization, requires the basic parameters to be calculated directly. In this situation, a logic consequence was the demand that the measuring system, in addition to the capability of higher time accuracy of records, might be used for primary computations. These computations permit, in case of optimization studies in particular, the experiments to be carried out in a programmed manner, i.e. a "measurement — computations—interpretation—measurement" cycle. In a so-organized cycle, the optimization effect can be reached by a considerably reduced number of experiments and with an effective use of the waiting times between the experiments.

The requirements set above have been used as the basic for the assumptions developed for the specialistic measuring system which had to make one of the links of the entire system planned. The system accomplishment, as described, assumed additionally a partial use of the already possessed and operationally proved equipment blocks. Such system, according to the our above general lines, has been built at the Institute of Nuclear Research (Nuclear Electronics Dept.) in form of a single-crate set of CAMAC modules. Making up this set are among others: autonomous processor (type 131), semiconductor memory modules, and peripherals. The set configuration and software make the system functionally suitable to be used as a multichannel programmed analyser or minicomputer. The set has a relatively broad for this type and size of equipment (peripheral outfit) including that obtained partly within the aid by International Atomic Energy Agency (which extends the set functional versatility).

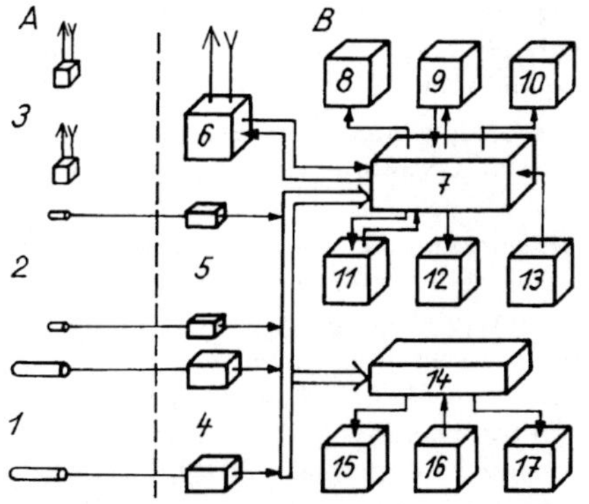

Fig. 3. Measuring system functionally-intended configuration schematic; A – detector set: 1 – probes with cable connections; 2 – physical-chemical parameters sensors; 3 – probes with logicsm memory modules and trasmit/receive units

B – measuring system: 4 – set of power supply modules; 5 – set of A/D converters; 6 – transmit/receive unit; 7 – CAMAC setup; 8 – screen display; 9 – printer; 10 – tape puncher; 11 – tape reader; 12 – X-Y recorder; 13 – magnetic disk store; 14 – IZOLDA setup; 15 – printer; 16 – tape reader; 17 – tape puncher

In its basic configuration, the Camac set as a multichannel analyser permits simultaneous recording from 16 measuring circuits with a minimum time per channel of 0·001 s. The capacity of a single measuring circuit is 512 channels with a possibility of setting three different time/channel values. This feature, with a limited number of channels, permits the measurements time to by greatly extended, the assumed accuracy being yet preserved.

The standard routines permit rapid operations and technological parameters calculations being a groove for this type of measurements.

The system has been checked for several years of practice and testified to be in compliance with the assumptions. Now, according to the accepted realization procedure, the system is being extended up to its intended configuration (by complementing with magnetic cassette memory, remote-measurements link module, and a set of A/D converters).

The complete intended configuration of this system is schematically shown in Fig. 3. The consequence of the step-by-step extension of this system is the benefit of keeping to it the proved units of the "standard" equipment (LT and HT power supplies, amplifiers, and discriminators) only partly modified. The system will also include the IZOLDA set described above. Operationally, the so-organized Standard-CAMAC measuring system will perform the following three basic functions:

– multichannel analyser for spectrum analysis: fitting of operational parameters (EHT, gain, discrimination thresholds) for individual detectors, not only to enable the measurement stability to be preserved but also to provide a possibility of separating the energy bands. This possibility is particularly essential in case of simultaneous tracing of, e.g. a number of different phases or components of tested medium by energetically different radioactive tracers;
– multichannel analyser operating as a multiscaler: gathering of data from 15 (IZOLDA) or 16 (CAMAC) measuring circuits according to the initially programmed parameters with the display (screen display) and recording (magnetic tape, punched tape, printer, X-Y recorder) of these data. The A/D converters permit, moreover, the parallel recording of technological parameters (e.g. temperature, pressure, flow rate, density, etc). The correlation of these parameters to the measurements of radioactive tracer concentrations will yield information being more complete description of the investigated process and allowing more correct interpretation and drawing more discerning conclusions;
– minicomputer to perform routine operations according to standard routines on the sets of results obtained (background substraction, allowance for decay, etc.), conversions (normalization, computation for decay, etc.), conversions (normalization, computation of integral curves, etc.), and computation of basic technological parameters (flow velocities, mean residence times, homogenization ratio, etc.). There's also a possibility of more extended use of the minicomputer by employing any software for calculations, including simple modelling of tested objects.

The realized project provides also for a functionally new solution consisting in the use of wireless telemetric communication between the probes and measuring apparatus. This problem is now inrealization stage and the project provides for the use of battery-operated detectors fitted with logic circuitry including memory modules, and radio transmitting-receiving units. Such a solution will alow the measuring system to be used in the studies of large objects also, thus, in cases where the distances between individual measurement points and field conditions prevent the use of conventional cable links. Such situation exists, for instance, in the investigation of industrial waste water installations, pipeline networks of heat and power plants, and hydrodynamic processes in open waters.

The final effect should be a possibility to carry out many hours measurements under various conditions with measuring point being even 2 or 3 km away from measuring van.

The project developed assumes that this solution will be functionally fully interchangeable with the wire communication system used in the present. This interchangeability allows also the measuring system operation in a hybrid configuration, i. e. using a part of conventional detectors and other part of detectors operating wirelessly. Moreover, an initial getting run of the logic of these circuits indicates that the above – described equipment setup with autonomous probes theoretically permits the planning of experiments with simultaneous recording at up to 115 measuring points.

Concurrently with the transmission system project presented work is proceeded on the replacement of scintillation detectors used to the present by semiconductor detectors of special configuration.

Mobile measuring laboratory

Within the complex of questions making up the problem of functionality of measuring systems intended for industrial tracer measurements, the conditions under which the equipment is to be used are equally essential question as the equipment itself. The substance of this problem is associated with the requirement for the equipment operational reliability, hence, with the possibility of carrying out the experiments as planned.

Unfortunately, only laboratory-grade measuring setups are capable of coping with the functional requirement described above.

The first factor that selects the type of equipment is its ability to be transported to measurement site and to

a) b)

Fig. 4. Parts of measuring van interior:
a) cabined accomodating CAMAC and IZOLDA setup; b) operator's stand

operate under severe industrial conditions. A tipical example is magnetic disk store the use of which under such conditions is hardly conceivable. This problem still presents a substantial, though no so drastic, difficulty in case of such apparatus as Standard, CAMAC and peripherals.

We settleted by us as a general rule that since the equipment to be used is laboratory-grade apparatus, then necessary to provide the operating conditions as close to the right ones as possible.

This statement has been laid down as the basis for the decision to build a mobile laboratory for industrial tracer measurements. The design assumptions, technical documentation, and practically all the workmanship have been accomplished at the Department of Nuclear Methods for Process Engineering. The laboratory equipment has been accomodated in an adequately modified AUTOSAN motor coach. All the coachwork has been (after closing useless windows down) lined with thermal insulation. The van interior has been divided into two functionally different compartments: equipment compartment and crew compartment. The equipment compartment has been fitted with ventilation (filtered air blow) and heating installations, and reasonably air-tightened. The limited space required a special arrangement of the measuring equipment in two separate cabinets provided with special shock absorbers and additional air tightening. The peripheral devices have been installed in similar manner.

The system operator's stand includes a teletype (for communication with processor and for experiment run control), screen display, and X-Y recorder. The equipment compartment interior architecture has been designed for the Standard-CAMAC system described before, and intended for permanent installation of the system.

The entire laboratory is powered from external electric mains, and the internal connection arrangement permits phase separation for the equipment supply. This supply circuit has been provided with an interference filter and, additionally a low-voltage supply circuit (from the van battery) installed to power the memory modules in case of mains voltage failure.

Some fragments of the laboratory described are shown in Fig. 4a and 4b.

The other modifications related to the trunk compartments (accomodating cable drums, electric devices for cable reeling, and water containers) and to the organization of crew compartment intended for transport, work and rest of the measuring staff members.

During several years of use, the laboratory van has proved to be functional enough for the needs of the measurments under industrial conditions which confirms the project assumptions.

Recapitulation

The complex of problems presented above concerns the realization of universal measuring system intended for industrial and field investigations of dynamic phenomena. A substantial part of these problems as realized and verifield in practice allows the statement that the accomplishment of such system is practicable.

The novelty of the designed system is, on one hand, its specialistic application for definite-type studies, and, on the other, its functional versatility. This versatility, apart from the technical facilities of using radioactive tracers for the investigations of any dynamic processes on any objects, consists also in the capability of parallel recording of basic process parameters. The coming-out beyond the conventional data collecting derived from radiactive tracer only is also an attenpt to obtain a more comprehensive representation of investigated phenomena which brings the general aim, i.e. the precise cognizance of these phenomena, nearer.

Received June 20, 1984

Isotopenpraxis **21** (1985) 5, S. 180 bis 183

Thermochromatographische Bestimmung der Adsorptionsenthalpie von Astat und Radon auf Palladiumoberfläche

B. Eichler, K. Son Chun

(Vereinigtes Institut für Kernforschung Dubna) [1])

Zur Untersuchung der Adsorption von Astat und Radon an Palladiumoberfläche wurden thermochromatographische on-line und off-line-Experimente mit den Radionukliden ^{210}At und ^{220}Rn durchgeführt. Die partiellen molaren Adsorptionsenthalpien für Nullbedeckung wurden experimentell bestimmt zu: $\Delta\bar{H}^0_{a,\,loc.}(At) = -(158 \pm 10)$ kJ Mol^{-1} und $\Delta\bar{H}^0_{a,\,mob.}(Rn) = -(37 \pm 4)$ kJ Mol^{-1}. Die Ergebnisse wurden mit theoretischen und experimentellen Werten anderer Elemente der sechsten Periode verglichen. Das Adsorptionsverhalten des Astats an Palladiumoberfläche entspricht dem der p-Metalle.

In order to investigate the adsorption of astatine and radon on a palladium surface some on- and off-line thermochromatographic experiments were carried out with ^{210}At- and ^{220}Rn-tracers. The partial molar adsorption enthalpy for zero covering was experimentally found to be $\Delta\bar{H}^0_{a,\,loc.}(At) = -(158 \pm 10)$ kJ mole^{-1} and $\Delta\bar{H}^0_{a,\,mob.}(Rn) = -(37 \pm 4)$ kJ mole^{-1}. The results are compared with the theoretical and experimental values for other elements of the sixth period. The adsorption behaviour of At is in conformity with that of the p-metals on a palladium surface.

Keywords

adsorption; adsorption heat; astatine 210; palladium; radon 220; surfaces; thermochromatography

Einleitung

Das Adsorptions- bzw. Desorptionsverhalten von Astat auf Metalloberflächen wurde von verschiedenen Autoren experimentell untersucht [1–3]. In einer Silberkolonne wurde erstmalig eine thermochromatographische Trennung des Astats von Jod und Quecksilber realisiert [2]. Thermodynamische Daten zur Charakterisierung des Adsorptionsverhaltens von Astat auf Metalloberflächen liegen nicht vor.

In der vorliegenden Arbeit werden die partielle molare Adsorptionsenthalpie der Radionuklide ^{210}At und ^{220}Rn auf Palladiumoberfläche in Gegenwart von Wasserstoff (Trägergas) bestimmt.

Experimente

Der Aufbau der Apparatur für thermochromatographische Untersuchungen wurde früher mehrfach beschrieben (z.B. [4]). Durch Kühlung des Kolonnenendes mit flüssigem Stickstoff konnten mit der verwendeten Apparatur thermochromatographische Experimente im Temperaturbereich zwischen einer Starttemperatur $T_s = 1100$ K und 140 K mit einem Temperaturgradienten von $g = -40$ K $\times$ cm^{-1} durchgeführt werden. Als Trägergas wurde Elektrolytwasserstoff mit einer Strömungsgeschwindigkeit von $v_0 =$ 0,28 cm^3 s^{-1} verwendet.

Die eingesetzten Quarzkolonnen wurden mit Palladiumfoliestreifen ausgelegt, deren geometrische Oberfläche $a_1 = 0{,}4$ cm^2 pro 1 cm Kolonnenlänge betrug. Die ^{210}At-Proben wurden durch Trocknung einer Lösung auf Palladiumfolie hergestellt und in dieser Form in der Kolonne (bei T_s) angebracht.

Die Experimente mit ^{220}Rn wurden im on-line Regime durchgeführt. In der Startposition der Kolonne war eine ^{228}Th-Quelle positioniert, die die erforderliche ^{220}Rn-Aktivität ($T_{1/2} = 55{,}6$ s) lieferte. Als Versuchsdauer wurde in diesem Falle die mittlere Lebensdauer dieses Nuklides angenommen. Die Abscheidungszone des Radon wurde durch die nichtflüchtigen Zerfallsprodukte ^{212}Pb und ^{212}Bi markiert. Nach Versuchsabschluß wurden die Palladiumfolien in 5 mm lange Meßproben zerschnitten und gammaspektrometrisch die Position des Maximums der Aktivitätsverteilung (Abscheidungstemperatur T_A) bestimmt.

Grundlagen

Die zur Bestimmung der Adsorptionsenthalpien $\Delta\bar{H}^0_a$ aus thermochromatographischen Daten erforderlichen Beziehungen wurden in [5] abgeleitet. Zur Interpretation thermochromatographischer Versuchsergebnisse kann sowohl ein Modell der mobilen (mob.) als auch ein Modell der lokalisierten Adsorption (loc.) herangezogen werden. Der Zusammenhang zwischen den thermodynamischen Standardgrößen der Adsorption, den experimentellen Parametern und der Abscheidungstemperatur ist für beide Modellfälle nach [5] gegeben mit:

$$-\frac{t_v \bar{v}_0 g}{a_1 T_0 \frac{V}{A} \exp\left(\frac{\Delta\bar{S}^0_{a,\,mob.}}{R}\right)} = Ei^*\left(\frac{-\Delta\bar{H}^0_{a,\,mob.}}{RT_A}\right) - Ei^*\left(\frac{-\Delta\bar{H}^0_{a,\,mob.}}{RT_s}\right) \quad (1)$$

und

$$-\frac{t_v \bar{v}_0 g}{a_1 T_0 \frac{V}{A(1-\Theta^\circ)} \exp\left(\frac{\Delta\bar{S}^0_{a,\,loc.}}{R}\right)} = Ei^*\left(\frac{-\Delta\bar{H}^0_{a,loc.}}{RT_A}\right) - Ei^*\left(\frac{-\Delta\bar{H}^0_{a,\,loc.}}{RT_s}\right) \quad (2)$$

[1]) Anschrift: P. O. Box 79, Moskau, UdSSR

Dabei bedeuten:

t_v die Versuchsdauer, s
$\bar{v}_0$ den Gasdurchsatz unter Normalbedingungen, $cm^3\,s^{-1}$
g den Temperaturgradienten ($g < 0$), $K\,cm^{-1}$
a_1 die innere freie Oberfläche pro 1 cm Kolonnenlänge, cm
T_0 die Standardtemperatur
T_s die Starttemperatur
R die allgemeine Gaskonstante

Die partiellen molaren Standardadsorptionsentropien können nach [5] berechnet werden mit:

$$\Delta\bar{S}^0_{a,\,mob.} = R \ln\left(\frac{1}{1(cm)\,\nu^0_{Pd}}\left(\frac{kT}{2\pi m}\right)^{1/2}\right) + 1/2\,R \quad (3)$$

bzw.

$$\Delta\bar{S}^0_{a,\,loc.} = R \ln\left(\frac{1}{\nu^0_{Pd}}\left(\frac{N}{1(cm)\,A_m} - \frac{N}{V}\right)\left(\frac{kT}{2\pi m}\right)^{3/2}\right) + 3/2\,R \quad (4)$$

mit:

k *Boltzmann*-Konstante
m Molekülmasse
T absolute Temperatur (in den folgenden Rechnungen $= T_A$)
N *Loschmidt*sche Zahl
ν^0_{Pd} Grenzfrequenz des Pd ($5{,}5 \times 10^{12}\,s^{-1}$)

Als Standardzustände wurden definiert:
das Standardvolumen des gasförmigen Adsorbats

$$V = 2{,}48 \times 10^{10}\,cm^3$$

(Standardmolvolumen bei einem Druck von 1 dyn cm^{-2}), und die Standardmolfläche des Adsorbats

$$A = 2{,}48 \times 10^{10}\,cm^2\,.$$

Mit diesen Festlegungen gilt $\frac{V}{A} = 1$ cm.

Der Standardbedeckungsgrad $\Theta°$ wird als Verhältnis der Fläche eines Mols des Adsorbats im Zustand einer dicht gepackten Monoschicht A_m und der Standardmolfläche A definiert:

$$\Theta° = \frac{A_m}{A}$$

A_m kann näherungsweise als Funktion des Atomradius r_A (im Falle des At des Kovalenzradius $r_{At} = 1{,}45 \times 10^{-8}$ cm) berechnet werden:

$$A_m = 2\sqrt{3}\,N r_A^2$$

Als Bezugszustände wurden das ideale Gas und der Zustand der Nullbedeckung betrachtet.

Zur Vereinfachung der Auswertung können die Differenzen der Integral-Exponentialfunktion $Ei^*(x)$ (Tabellen und Näherungslösungen s. [5]) halblogarithmisch als Funktion der Abscheidungstemperatur T_A für eine Reihe von $\Delta\bar{H}^0_a$-Werten im Abstand von jeweils 10 kJ Mol^{-1} dargestellt werden. Auf diese Weise wurde aus den berechneten Werten der linken Gleichungsseiten (1) bzw. (2) sowie den gemessenen Abscheidungstemperaturen (T_A) $\Delta\bar{H}^0_a$ interpolativ bestimmt.

Ergebnisse und Diskussion

Die experimentell bestimmten Abscheidungstemperaturen des At und Rn auf Pd sind in Tab. 1 zusammengestellt. Da

Tab. 1. Versuchsbedingungen und Ergebnisse thermochromatographischer Experimente mit ^{210}At und ^{220}Rn; thermodynamische Standardgrößen der Adsorption auf Palladiumoberfläche

Adsorbat: ^{210}At		Versuchsregime: off-line	
Versuchsdauer	Abscheidungstemperatur	$\Delta\bar{S}^0_{a,\,loc.}$	$\Delta\bar{H}^0_{a,\,loc}$
s	K	$JMol^{-1}\,K^{-1}$	$kJ\,Mol^{-1}$
3600	528	−213,6	−153,3
1600	603	−211,9	−173,5
5400	515	−213,9	−152,5
3600	515	−213,9	−150,7
		∅:	−158 ± 10
Adsorbat: ^{220}Rn		Versuchsregime: on-life	
Versuchsdauer $\frac{T_{1/2}}{\ln 2}$	Abscheidungstemperatur	$\Delta\bar{S}^0_{a,\,mob.}$	$\Delta\bar{H}^0_{a,\,mob.}$
s	K	$JMol^{-1}\,K^{-1}$	$kJ\,Mol^{-1}$
80,2	168 ± 10	−172,0 ± 0,1	−37 ± 4

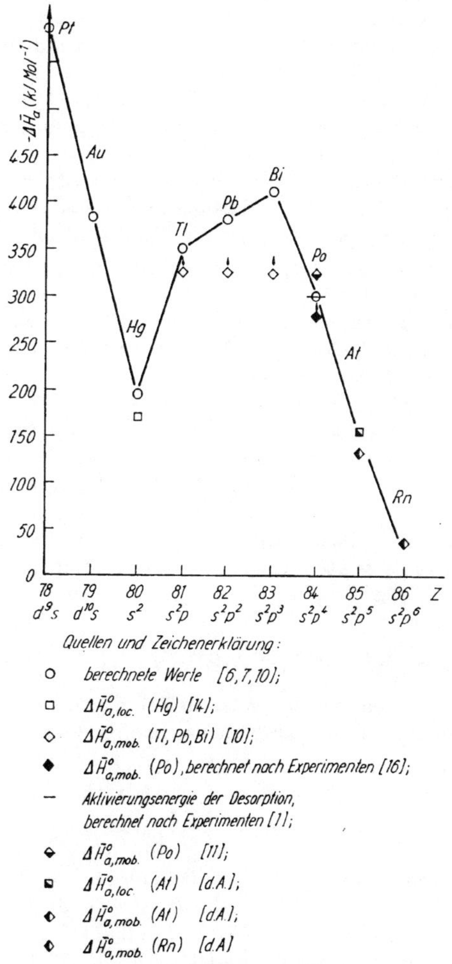

Abb. 1. Berechnete und experimentell bestimmte Adsorptionsenthalpien einiger Elemente der 6. Periode auf Palladiumoberfläche als Funktion der Ordnungszahl Z bzw. der Elektronenkonfiguration im Grundzustand

Tab. 2. Standardenthalpien der Komponenten des Systems Wasserstoff – Halogen – Palladium sowie Kovalenzradien der Halogene [12]

Halogen X =	F	Cl	Br	J	At
Standardenthalpie des gasförmigen atomaren Halogens X ΔH^0_{298} X (g), kJ g-Atom^{-1}	79,5	121,3	111,8	106,8	(100)
atomare Bildungsenthalpie der gasförmigen Verbindung HX $\Delta H^*_{298;f}$ HX (g), kJ Mol^{-1}	−568,5	−431,6	−366,1	−298,3	(−180)
Standardbildungsenhtalpie der gasförmigen Verbindung HX $\Delta H^0_{298;f}$ HX (g), kJ Mol^{-1}	−271,0	−92,2	−36,3	+26,6	((+138))*
Standardbildungsenthalpie des festen Palladiumhalogenids $\Delta H^0_{298;f}$ PdX_2 (s), kJ Mol^{-1}	−485,2 [13]	−163,0	−120,0	−63,0	(+27)
Standardenthalpie der Reduktionsreaktion des PdX_2 mit Wasserstoff $\Delta H_{298;red.}$, kJ Formelumsatz^{-1}	−65,8	−21,4	+47,4	+116,2	(+249)*
Kovalenzradien, 10^{-10} m	0,64	0,99	1,14	1,33	1,45

( ): extrapoliert als Funktion der Ordnungszahl;
( )*: berechnet mit extrapolierten Werten.

keine theoretisch berechneten oder experimentellen Vergleichsdaten [17, 18] vorliegen, ziehen wir zum Vergleich den theoretischen Verlauf der $\Delta\bar{H}^0_a$-Werte [6, 7] innerhalb der 6. Periode in Abhängigkeit von der Ordnungszahl (bzw. Elektronenkonfiguration) heran. In Abb. 1 werden die berechneten $\Delta\bar{H}^0_a$-Werte mit verfügbaren experimentellen Daten verglichen. Für Tl, Pb und Bi liegen nur untere Grenzwerte vor [10].

In der Wechselwirkung von Metallen mit Palladium spielt der Elektronentransfer zum nicht aufgefüllten 4d-Band des Pd die dominierende Rolle. Der Kurvenverlauf demonstriert deutlich den Einfluß der Schalenabschlüsse bei Au, Hg und Rn und die Wirkung freier nicht gekoppelter p-Elektronen. In diesen Verlauf fügen sich die experimentell bestimmten $\Delta\bar{H}^0_a$-Werte für At und Rn widerspruchsfrei ein.

Infolge der Ko-Adsorption von Wasserstoff – die partielle molare Adsorptionsenthalpie des Wasserstoffs bei Nullbedeckung beträgt −117 kJ Mol^{-1} [9] – stellen die erhaltenen Werte wahrscheinlich untere Grenzwerte dar.

Mit hoher Wahrscheinlichkeit wird die experimentell bestimmte Adsorptionsenthalpie des At durch die Wechselwirkung von Astatatomen mit der Palladiumoberfläche realisiert.

Als Alternative dazu wird versucht, den beobachteten Adsorptionsprozess als dissoziative Chemisorption von HAt zu beschreiben (d. h. $\Delta\bar{H}^0_{a,\,loc} = \Delta\bar{H}^0_{a,\,diss.}$).

Mit

$$\Delta\bar{H}^0_{a,\,diss.} = -\Delta H^*_{298;f}\mathrm{HAt}(g) + \Delta\bar{H}^0_a H + \Delta\bar{H}^0_a\mathrm{At} \quad (5)$$

folgt unter Verwendung der Daten in Tab. 2 und einer nach [9] berechneten Standardadsorptionsenthalpie des atomaren Wasserstoffs $\Delta\bar{H}^0_a\mathrm{H} = -276{,}5$ kJ g-atom^{-1} für die Standardadsorptionsenthalpie des atomaren At $\Delta\bar{H}^0_a$ At $= -61{,}5$ kJ g-atom^{-1}.

Verglichen mit Rn (s. Abb. 1) erscheint dieser Wert viel zu klein. Auch in Anbetracht der extrapolierten Enthalpiewerte des At und seiner Verbindungen (s. Tab. 2) liegt die Schlußfolgerung nahe, daß im Experiment die Adsorption von At-Atomen am Palladium erfolgt.

In diesem Zusammenhang ist ein Vergleich mit dem Adsorptionsverhalten von 34m-Cl auf Palladium, das in einer früheren Arbeit [14] unter vergleichbaren Bedingungen untersucht wurden, aufschlußreich. Aus diesem Experiment folgt

$$\Delta\bar{H}^0_{a,\,loc.}\,(34\text{m-Cl}) = -107 \pm 10 \text{ kJ g-atom}^{-1}\,.$$

Unter Annahme dissoziativer Chemisorption des mit Sicherheit vorliegenden chemischen Zustandes HCl (g) folgt für die Standardenthalpie der Adsorption des atomaren Cl auf Palladiumoberfläche mit Gl. (5)

$$\Delta\bar{H}^0_a\,\mathrm{Cl} = -262 \text{ kJ g-atom}^{-1}\,.$$

Während für atomare Halogene offensichtlich die Adsorptionsenthalpie auf Pd mit steigendem Z absolut abnimmt (analoges Verhalten wurde bereits in der früheren Arbeit [2] für Ag-Oberflächen gefunden), wird diesem Effekt im Wasserstoff-Trägergas die abnehmende Stabilität der Wasserstoffverbindungen der Halogene überlagert. Auf diese Weise ist die niedrige Adsorptionsenthalpie des (34m-Cl) gegenüber ^{210}At zu erklären, die einem dissoziativen Chemisorptionsprozess von HCl zugeordnet werden kann.

In [10] waren empirische Kriterien für das Vorliegen lokalisierter Adsorption bestimmt worden. Dafür war ein Mindestwert der Lösungsenthalpie im festen Zustand erforderlich (< +50 kJ Mol^{-1}) und ein Verhältnis von Abscheidungstemperatur zu Schmelztemperatur des Adsorbenten von <0,37. Diese Bedingungen sind für die Astat- und Chloradsorption (s. Tab. 2) erfüllt. Deshalb wird für diese Elemente lokalisierte Adsorption und im Falle des Radon mobile Adsorption angenommen.

Die Adsorptionsenthalpie des At unter Annahme eines mobilen Adsorptionsmodells wurde nur zum Vergleich in Abb. 1 mit angegeben. Die Adsorptionsenthalpie des Rn auf Pd (s. Tab. 1) entspricht den typischen Werten für seine Adsorption auf Festkörperoberflächen: an Kohle −31,4 kJ Mol^{-1}, an Silikagel −37,7 kJ Mol^{-1} [15].

Der experimentell für Pd gefundene Wert übertrifft die Verdampfungswärme des Rn [12] um den Faktor 2,2. Die Radon-Adsorptionsenthalpie ist offenbar weitgehend adsorbentunabhängig.

Die Abfolge der berechneten sowie experimentell bestimmten Adsorptionsenthalpien der Nichtübergangselemente der 6. Periode zeigt die Möglichkeit hocheffektiver Abtrennung des Astats insbesondere von den Elementen Bi und Po.

In seinem Adsorptionsverhalten auf Pd-Oberflächen entspricht das Astat einem leichtflüchtigen p-Metall mit einem ungepaarten Elektron.

Eingegangen am 4. 9. 1984

Literatur

[1] *G. Johnson, R. Leiniger, E. Segre,* J. chem. Phys. **17** (1949) 1
[2] *B. Eichler, V. P. Domanov,* Preprint JINR P-12-7928, Dubna 1974
[3] *V. M. Wachtel, G. V. Vinel, Z. Vylov, I. I. Gromova, A. V. Novgorodov, Ju. V. Norseev, V. A. Khalkin,* Isotopenpraxis **12** (1976) 441
[4] *B. Eichler, G. Loos, M. Saupe,* Report ZfK-422, Rossendorf 1980
[5] *B. Eichler, I. Zvara,* Radiochimica Acta **30** (1982) 233
[6] *B. Eichler,* Report ZfK-396, Rossendorf 1979
[7] *H. Roßbach, B. Eichler,* Report ZfK-527, Rossendorf 1984
[8] *A. W. Bryant, W. G. Bugden, J. N. Pratt,* Acta Metallurgica **18** (1970) 101
[9] *J. R. Anderson,* "Structure of Metallic Catalysts". London: Academic Press 1975
[10] *B. Eichler, S. Hübener, B. L. Zhuikov, H. Roßbach,* Preprint JINR P-12-81-717, Dubna 1981
[11] *B. Eichler, S. Hübener, H. Gäggeler, H. Roßbach,* Radiochimica Acta, in press 1985
[12] *A. I. Jefimov,* "Svoistva neorganičeskich soedinienij". Leningrad: Izd. Khimija 1983
[13] *M. Ch. Karapetjanz,* "Osnovnye termodinamičeskye konstanty neorganičeskich i organičeskich veščestv". Moskva: Izd. Khimija 1968
[14] *B. Eichler, Kim Son Chun,* Preprint JINR P12-83-206, Dubna 1983
[15] *K. Bagnall,* "Chemistry of the Rare Radioelements Po-Ac". London: Burtherworth Scientific publications 1957
[16] *N. Greulich, H. Dornhöfer, W. Schmidt-Ott, H. Gäggeler, W. Weber,* Report GSI 82-1, Darmstadt 1982, p. 218
[17] *E. H. Appleman,* "The Radiochemistry of Astatine". NAS-NS-3012, 1960
[18] *A. H. W. Aten,* jun., *T. Doorgeest, U. Hollstein, H. P. Moeken,* The Analyst **77** (1952) 774

Isotopenpraxis **21** (1985) 5, S. 183 bis 186

Determination of Isotope Ratios and Concentrations of Uranium and Plutonium in Nuclear Fuels

A. Hermann, U. Niese, H. Stephan, S. Hübener, S. Niese

(Academy of Sciences of the GDR, Central Institute of Nuclear Research, Rossendorf)[1]

Methods of separation of uranium and plutonium from irradiated nuclear fuels by extraction and ion exchange are described as well as methods of the determination of the element concentration or isotopic composition by spectrophotometry, coulometry, potentiometric titration, mass spectrometry, alpha spectroscopy and liquid scintillation counting. Advantages and disadvantages of the methods are discussed.

Methoden für die Abtrennung von Uranium und Plutonium aus bestrahlten Kernbrennstoffen durch Extraktion und Ionenaustausch sowie für die Bestimmung der Konzentrationen und der Isotopenzusammensetzung durch Spektrophotometrie, Coulometrie, Potentiometrische Titration, Massenspektrometrie, Alphaspektrometrie und Flüssigszintillationszählung werden vorgestellt. Die Vor- und Nachteile der Methoden werden diskutiert.

Keywords

alpha spectroscopy; isotope ratio; mass spectroscopy; nuclear fuels; plutonium; potentiometry; quantitative chemical analysis; uranium; voltametry

1. Introduction

Methods of uranium and plutonium analysis have been developed for the burnup determination of spent nuclear fuels [1–3]. Other methods are more convenient for unirradiated fuel, e.g. the potentiometric titration of uranium. In general in the case of spent fuel a separation process will become necessary in order to get the pure fuel components. Especially in safeguards analysis a high accuracy in determining fissile nuclides is demanded. The methods developed should be briefly summarized in this paper followed by a discussion of the advantages and disadvantages of the selected methods.

2. Experimental

2.1. Separation methods

Extraction. In cases when only one component, uranium or plutonium has to be separated from the fuel solution simple extraction processes are sufficient. Tributylphosphate (TBP) or Trioctylphospine oxide (TOPO) in organic diluents are used as extractants for uranium.

With 20% TBP in n-heptane uranium is extracted from 3 N HNO_3 solution. Remaining fission products (FP) are removed from the organic phase by four HNO_3 washings. The reextraction is accomplished with water (four times) [4]. Decontamination from Pu is achieved by its reduction with 0·2 M hydrazine in 3 N HNO_3 prior to the extraction. Usually decontamination factors of 10^4-10^5 from gross FP activity have been reached. This method has been applied for isotope or isotope dilution analysis of U, where a reduced chemical yield is acceptable.

0·1 M TOPO in cyclohexane quantitatively extracts uranium from 1 N HNO_3. After washing with 1 N HNO_3 the reextraction is performed using 0·2 M Na_2CO_3 solution (two times). The reextract is acidified in accordance with the method of determination [5]. Here too the reduction of Pu is necessary in order to get a pure U fraction. This procedure is convenient for the quantitative determination of uranium by coulometry or spectrophotometry.

Plutonium is purified by thenoyltrifluoroacetone (TTA) extraction of Pu(IV) [6]. The sample solution is conditioned

[1]) Anschrift: DDR-8051 Dresden, PF 19

by hydroxylamine (→ Pu-III) and $NaNO_2$ (→ Pu-IV) in order to get an aqueous phase of Pu(IV) in 1 N HNO_3 suitable to the extraction with 0·5 M TTA in xylole. After several washings with 1 N HNO_3 reextraction with 10 N HNO_3 results in a pure plutonium fraction suitable for alpha- or mass-spectrometry.

Chromatography. With the aid of chromatographic column methods high decontamination factors ($>10^4$) have been achieved for both uranium and plutonium.

An extraction chromatographic method uses 100% TBP on polytetrafluoroethylene powder as stationary and HNO_3 solutions of various composition as mobile phases for separating U and Pu as purified fractions from FP (and Am/Cm) [7]. After washing out the FP (and Am/Cm) with 5 N HNO_3 Pu is removed by reducing elution using a solution of 1·8 N HNO_3, 0·1 N H_2SO_3 and 0·05 M HSO_3NH_2. Then uranium is eluted with water. A quantitative recocovery of both U and Pu with decontamination factors from FP of $\approx 10^7$ (for U) and $\approx 5 \cdot 10^4$ (for Pu) and a high degree of purification from each other has been observed.

Another method in use is the anion exchange chromatography on resins like Dowex 1 × 8 [8, 9]. For conditioning to Pu(IV) and isotope exchange in the case of added spike solution reduction with $FeSO_4$ solution followed by oxidation with concentrated HNO_3 are adequate measures. Sorption of U and Pu as anionic nitrate complexes and elution of FP are performed with 7·5 N HNO_3 (4 ml). With the same acid U is eluted from the column and with 0·35 N HNO_3 the remaining Pu is removed. The purity of the separated U meets the demands of mass-spectrometry while that of Pu depends on the elution volume of U [10].

2.2. *Spectrophotometry*

The reagents arsenazo III (AA III) and dibenzoylmethane (DBM) are used in the spectrophotometric determination of uranium [1, 3, 5]. In fuel solutions of appropriate composition U has been determined immediately in the 9 N HNO_3 being 0·03% in AA III by direct spectrophotometry at $\lambda = 655$ nm. If necessary TOPO extraction is adopted to this method which allows a content as low as 15 µg U to be determined. The DBM method is not so sensitive but offers greater stability of the coloured solution developed with pyridine buffer in alcohol — water — mixture. Here, too, direct spectrophotometry at $\lambda = 405$ nm is applied. Results on fuel solutions of the WWER-70 reactor are given in comparison with coulometric ones in Tab. 1.

Tab. 1. Results of uranium determinations in spent nuclear fuel solutions, mg U/g soln.

Sample-No.	Spectro-photometry	Coulometry	% Difference
1	2·95 ± 0·03	2·92 ± 0·02	+1·0
2	3·24 ± 0·04	3·22 ± 0·02	+0·6
3	4·91 ± 0·04	4·87 ± 0·03	+0·8
4	5·60 ± 0·08	5·54 ± 0·05	+1·1
5	4·38 ± 0·04	4·35 ± 0·02	+0·7
6	4·67 ± 0·07	4·70 ± 0·05	−0·6

Plutonium has been determined with AA III after its extraction chromatographic separation from U and FP [11]. The Pu fraction is treated with $NaNO_2$, an excess of which is destroyed with urea. After bringing concentrations of HNO_3 to 5−7 N and of AA III to 0·0012 M extinction at $\lambda = 665$ nm is measured against a solution of the reagents. A standard deviation of ±3% has been established for this method.

2.3. *Coulometry of Uranium*

Quantities of 2−10 mg U have been determined by potentiostatic coulometry measuring the electricity necessary for the reduction of UO_2^{2+} to U(IV) [2, 3, 5]. The current flowing through the electrolytic cell consists of two components due to the electrochemical reaction and other redox processes giving cause to the so called background current. The latter one has to be thoroughly determined in order to get a high precision [12]. The two components can be described by the equation

$$Q_t = Q(1 - e^{-kt}) + k_b \cdot t \qquad (1)$$

Q_t, Q — electricity at time t and that responsible for the analytical reaction; k, k_b — velocity constants of the analytical and the background reactions respectively.

As equipment an electrolytic cell with mercury on the bottom as working electrode and a Pt wire in a diaphragma as second electrode has been connected with a coulometer OH 404 (Radelkis, Hungary). The experimental data were treated according to equation (1) by a computer using the programme "nilireg". Tab. 1 gives examples of results on WWER fuel solutions.

2.4. *Potentiometric Titration of Uranium*

For precise uranium determinations the method of *Davies-Gray-Eberle-Lerner* [13−16] is used. Uranium is reduced in concentrated phosphoric acid containing amidosulfonic acid with $FeSO_4$ and an excess of which is treated in a Mo(VI) catalysed HNO_3 oxidation step. In the following titration with $K_2Cr_2O_7$ solution vanadyl sulphate serves as a catalyst. Results of uranium determinations in the uranium oxide powder SR-50, distributed by the IAEA Analytical Quality Control Service, are given in Tab. 2. Our overall mean of 86·097% U differes by −0·145% U from the reference value of 86·242% U. This difference is statistically not significant at the 95% probability level.

Tab. 2. Results of the potentiometric titration of uranium in IAEA standard sample SR-50 (uranium oxide powder), % U

Aliquot-No.	Dissolution-No.		
	1	2	3
1	86·056	86·111	85·944
2	86·257	86·057	85·971
3	86·365	85·896	86·218
mean	86·226	86·021	86·044
SD	0·156	0·111	0·150
overall mean	86·097 ± 0·147		

2.5. *Mass-spectrometric Isotope Dilution*

The isotopic composition of U and Pu is measured with a TH-5 Varian MAT mass spectrometer additional connected with a device for automatically reaching the measurement conditions within a given time interval. Measurements have been performed on thoroughly separated fractions as well as on samples with approximately equal quantities of both U and Pu [17]. In the latter case the different

Tab. 3. Results of mass-spectrometric isotope and isotope dilution analysis of spent nuclear fuel samples, % atoms

Isotope	Sample-No. 1	2	3
^{234}U	0·0270 ± 0·0010	0·0245 ± 0·0012	0·0210 ± 0·0007
^{235}U	1·7322 ± 0·0033	1·3931 ± 0·0031	0·8832 ± 0·0025
^{236}U	0·3683 ± 0·0021	0·4251 ± 0·0017	0·5150 ± 0·0031
^{238}U	97·873 ± 0·009	98·1569 ± 0·005	98·581 ± 0·007
^{236}Pu (· 10^5)		1·05 ± 0·3[1])	1·43 ± 0·31[1])
^{238}Pu	0·732 ± 0·007[1])	1·186 ± 0·012[1])	1·954 ± 0·020[1])
^{239}Pu	69·34 ± 0·06	63·23 ± 0·06	55·76 ± 0·05
^{240}Pu	18·43 ± 0·06	20·50 ± 0·06	22·82 ± 0·05
^{241}Pu	9·67 ± 0·05	12·03 ± 0·09	13·94 ± 0·06
^{242}Pu	1·82 ± 0·009	3·05 ± 0·018	5·52 ± 0·022
Pu/U (· 10^3) mass ratio	7·34 ± 0·16	8·84 ± 0·25	11·89 ± 0·17

[1]) alpha-spectrometry

Tab. 4. Example of an alpha-spectrometric isotope dilution analysis of plutonium in spent nuclear fuel

Method (detector)	Data treatment	Activity ratio (239 + 240)/238	Specific activity kBq/g	Isotope ratio	$^{239+240}Pu$ µg/g
Si(Li)	addition	1·7224 ± 0·0140 0·3957 ± 0·0031[1])	12·866	1·736	3·756 ± 0·032
Si(ANOX)	addition	1·7656 ± 0·0148 0·3950 ± 0·0030[1])	12·750	1·720	3·715 ± 0·032
Si(ANOX)	computer programme[2])	1·8949 0·4075[1])	13·007	1·740	3·7973
mass spectrometry				1·749	

[1]) samples with added ^{238}Pu spike
[2]) self consistent fit [21]

temperatures of volatilization of U and Pu have been used, analysing U isotopes at lower temperatures (current of the volatilization filament: 1·8 A) and Pu isotopes at higher ones (current: 2·4 A). Data treatment with the aid of a computer has been arranged. In the case of isotope dilution analysis enriched ^{233}U and ^{242}Pu samples of known concentration have been used. Results of the analysis of isotopic composition and content of U and Pu in fuel samples are given in Tab. 3.

2.6. *Isotope Dilution Alpha Spectroscopy of Plutonium*

The isotope dilution alpha spectroscopy analysis (IDAS) of Pu is based on measuring the activity ratio $^{238}Pu/^{239+240}Pu$ in comparison with that ratio after a known amount of ^{238}Pu has been added to the sample solution. Alpha spectrometry has been performed after TTA extraction and electrolytic deposition of Pu on steel discs using a Si(Li) drift detector (177 mm²; FWHM 25 keV for ^{210}Po) or an ANOX (Si(SB)) detector (12·6 mm²; FWHM 12 keV for ^{210}Po). For data treatment besides the simple addition of counts the computer programme "orion" [18] has been applied. The alpha spectrometric IDA of Pu is illustrated by the data of Tab. 4 showing the precision and comparing the different possibilities concerning measurements and data treatment.

Results of both mass- and alpha-spectrometric IDA have been checked in the interlab-comparison "SROK" [18, 19].

2.7. *Determination of Alpha Acctivity by Liquid Scintillation Spectroscopy*

Liquid scintillation spectroscopy offers nearly 100% efficiency in alpha counting and a simple sample preparation [20]. The organic extract with TBP or TOPO may be mixed directly with the toluene based scintillator (Diphenyloxazol and Dimethylphenyl oxazolyl benzene). In commercial equipments the energy resolution amounts to about 50 percent. Therefore we cannot distinguish between alpha emitters of similar energy.

We must also take into account, that the photon yield of alpha particles per absorbed energy unit is by a factor of ten lower then the photon yield of beta particles. The broad alpha peaks are mixed with continuous beta spectra from beta emitters (^{241}Pu, ^{234}Pa, fission products). We use the liquid scintillation spectroscopy for the measurement of the total alpha activity especially for the determination of chemical yields in developing separation procedures.

3. Discussion

Safeguards analytical measurements need extreme accuracy in the determination of the amounts of fissile materials and in burn-up determinations accurate activity ratios should be measured. Fuel quality assurance demands high accuracy in the uranium content of samples to be analysed.

Our experience in using different separation and determination methods allows the following rules to be formulated:

For uranium determination the electrochemical methods are more precise than the spectrophotometrical ones but need higher amounts of substances e.g. 10 to 100 mg. Mass spectrometric isotope dilution allows high precission and accuracy with only small amounts of uranium necessary.

For the determination of plutonium the isotope dilution mass spectrometry (IDMS) is precise, allows the determination of the isotopic composition and needs only small amounts of plutonium. Isotopic dilution alpha spectroscopy (IDAS) is more accurate in the determination of the ^{238}Pu content. ^{238}Pu has a small percentage in the mass spectrum but a high one in the alpha spectrum as a result of its high specific activity. The contamination of ^{238}Pu in IDMS by ^{238}U and in IDAS by ^{241}Am is detectable more easily in the second case.

The spectrophotometry of Pu is convenient for the determination of the total Pu content.

According to the low selectivity but simple sample preparation of LSC this method should be prefered in the development of separation methods.

Received September 12, 1984

References

[1] *A. Hermann, L. Baraniak, S. Hübener, I. Katzwinkel, R. Legler,* Kernenergie 18 (1975) 157

[2] *A. Hermann, H. Stephan, S. Hübener, D. Nebel, S. Trebeljahr, U. Niese, F. Sus, E. Klosova,* Kernenergie **23** (1980) 177

[3] *A. Hermann, L. Baraniak, S. Hübener, D. Nebel, U. Niese, S. Trebeljahr,* Isotopenpraxis **12** (1976) 277

[4] *A. Hermann, H. Zahn,* ZfK-294 (1975) 15

[5] *A. Hermann, L. Baraniak,* ZfK-257 (1973)

[6] *U. Niese, H. Stephan, A. Hermann,* ZfK-451 (1981) 123

[7] *S. Hübener, A. Hermann,* J. Radioanal. Chem. **21** (1974) 157

[8] *I. K. Kressin, G. R. Waterbury,* Anal. Chem. **34** (1962) 1598

[9] *B. P. Freeman, J. R. Weiss, C. E. Pietri,* Rev. Sci. Instrum. **52** (1981) 1252

[10] *L. Koch,* Radiochim. Acta **12** (1969) 161

[11] *D. Nebel, A. Hermann, S. Trebeljahr,* J. Radioanal. Chem. **28** (1975) 133

[12] *L. Meites, S. A. Moros,* Anal. Chem. **31** (1959) 23

[13] *W. Davies, W. Gray, K. C. McLeod,* Talanta **17** (1970) 937

[14] *A. Eberle, K. Lerner,* NBL-252 (1970)

[15] *C. J. Goldbeck, M. W. Lerner,* Anal. Chem. **44** (1972) 594

[16] *J. Slanina, F. Bakker, A. J. P. Groen, W. A. Lingesak,* Z. Anal. Chem. **289** (1978) 102

[17] *H. Stephan, A. Hermann, G. Bell,* ZfK-489 (1982) 175

[18] *U. Niese, J. von Borany, P. Urwank, S. Niese,* Isotopenpraxis **20** (1984) 000

[19] *A. Hermann, H. Stephan, G. Bell, S. Trebeljahr,* ZfK-367 (1978) 17

[20] *W. J. McDowell, D. T. Farrar, M. R. Billings,* Talanta **21** (1974) 1231

[21] *P. Urwank,* Nucl. Instr. and Meth. **203** (1982) 329

Isotopenpraxis 21 (1985) 5, S. 186 bis 192

Buchbesprechungen

Nuclear Data for Science and Technology
Proc. of the International Conference,
Antwerp, 6–10 Sept. 1982

Hrsg.: K. H. Böckhoff. Dordrecht, Boston, London: D. Reidel Publ. Comp. 1983. XVIII + 1054 S., zahlr. Abb. u. Tab., (engl./frz.,) geb. Dfl 275,–; US$ 120.–

Die Konferenz in Antwerpen ist die Fortsetzung einer Reihe, die in den 60er Jahren begann und inzwischen einen etwa jährlichen Turnus erreicht hat. Die 1982er Veranstaltung wurde von dem westeuropäischen CEC-Verband in Kooperation mit der IAEA und der OECD organisiert und vereinigte 270 Wissenschaftler aus 35 Ländern, darunter estmalig Teilnehmer aus der VR China. Als besonderer Vorzug der Proceedings ist zu werten, daß neben den 92 im Plenum gehaltenen Vorträgen auch alle 158 der erstmalig zugelassenen Poster-Beiträge in ausgezeichneter Druckqualität enthalten sind.

Der Band gliedert sich in 17 Kapitel (12 davon sind als Programm-Sektionen ausgewiesen), die Teilnehmer- und Autorenliste und in den diese Tagung betreffenden CINDA-Index. Der Gesamtinhalt entspricht bei einer bemerkenswerten Vielfalt der Beiträge genau dem Titel und behandelt damit in größerem Umfang als bei bisherigen Konferenzen technologisch bedeutsame Resultate bei der und Forderungen für die Kerndatengewinnung (sowohl für Druckwasserreaktoren und Flüssigmetallbrüter als auch für zukünftige Fusionsanlagen oder intensive Neutronenquellen). In der Abschlußsitzung wird dies u. a. dadurch ausgedrückt, daß erstmalig ein vollständiger Überblick für ein bestimmtes Nuklid (^{238}U) auf einer derartigen Konferenz zustande gekommen ist.

Im einzelnen wurden folgende Abschnitte zusammengestellt:

- Neutronendaten der Transactinide (speziell Bestandteile des Spaltreaktorcores; 15 Beiträge),
- Neutronendaten für Konstruktionsmaterialien, Kühlmittel und Abschirmungen bei Spaltungsreaktoren (16 Beiträge),
- Kerndaten für Spaltprodukte und für den Spaltreaktor-Kernbrennstoffkreislauf (29 Beiträge),
- Kerndaten für Fusionsreaktoren und intensive Neutronenquellen (20 Beiträge),
- Dosimetrie-Kerndaten (13 Beiträge), Neutronenstandarddaten (11 Beiträge),
- Verständnis der Kernreaktionen und Kernmodellbe-
- rechnungen (23 Beiträge),
- Neutronendaten-Analyse und Einschätzungen (23 Beiträge),
- Zugrunde liegende Physik und Stützdaten (30 Beiträge),
- Ausrüstungen, Instrumente und Methoden zur Neutronendatenmessung (24 Beiträge),
- Kerndaten für biomedizinische Anwendungen (8 Beiträge),
- Kerndaten für das Verständnis der stellaren Kernsynthese (5 Beiträge),
- Festkörper-, molekulare und atomare Wirkungen auf Kerndaten (4 Beiträge),
- Allgemein interessierende Fragen (8 Beiträge).

Damit spannt sich der Bogen von den traditionellen Fragen der Kerndatengewinnung und -einschätzung (Bereitstellung exakterer Daten, Klärung von Diskrepanzen, Benchmark-Einschätzungen, Vorstellung von Experimenten und Projekten usw.) über die Theorie der Reaktionsmechanismen (einschließlich der Auswirkungen auf die Groß-Unifizierte Theorie), die Atomhülleneinflüsse auf Kerndaten, die für die Ermittlung von Strahlenschäden in KKW-Materialien und biologischem Gewebe relevanten Kerndaten (einschließlich der Einflüsse auf Entsorgungstechnologien) bis hin zur Produktion von kurzlebigen neutronendefizitären Nukliden mit ihren weitreichenden Anwendungen in der medizinischen Organ-Diagnostik. Hervorzuheben sind die zu einzelnen Schwerpunkten gehaltenen Übersichts- bzw. Einführungsvorträge (z. B. zu Materialfragen von *Rowlands* et al., UKAEA (GB), zur Tritiumbrütung in Fusionsblankets von *Abdou*, ANL (USA) oder zu den mathematischen Methoden für die Einschätzung experimenteller Daten von *Kalachev* et al., MIFI (UdSSR)). Der Trend, für wissenschaftlich-technische Belange Daten in der erforderlichen Qualität (d. h. Details in Resonanzstrukturen, höhere Präzision, bessere Konsistenz von Datensätzen) bereitzustellen, ist weiterhin Schwerpunkt bei der Kerndatengewinnung und greift zunehmend auf die Datenanwendungen über, indem z. B. die Anwenderrechenprogramme kritisch hinsichtlich der Resultate bei identischen Datensätzen verglichen werden.

Der Konferenzband gibt trotz der Vielfalt und Details einen ausgezeichneten Überblick über Methoden, Aufwand, Anwendungen und Nutzen der Kerndatengewinnung und ist deshalb auch einem Leserkreis von Natur- und Ingenieurwissenschaftlern zu empfehlen, der nur mittelbar auf Kerndaten angewiesen ist. Bei dem Gesamtumfang des Buches hätten sicher auch die Diskussionsbeiträge untergebracht werden können. Die hervorragende Ausstattung entspricht dem vorauszusehenden Interesse, das diesem Band entgegengebracht werden dürfte.

D. Zappe

Solid State Dosimetry
Proceedings of the VIIth International Conference on Solid State Dosimetry held in Ottawa (Canada) September 27–30, 1983

T. Stoebe, E. P. Goldfinch (eds.). Ashford, Kent: Nuclear Technology Publishing 1983 (Radiation Protection Dosimetry Vol. 6) 370 S., div. Abb. u. Tab. (CONF – 830953); kart. $ 80.–

Die aus den Konferenzen zur Lumineszenzdosimetrie hervorgegangene Internationale Konferenz zur Festkörperdosimetrie fand zum siebenten Mal statt. An ihr nahmen 174 Wissenschaftler aus 24 Ländern teil, die insgesamt 106 Vorträge hielten. Die nunmehr vorliegenden Proceedings wurden sehr schnell nach der Konferenz publiziert.

Entsprechend der auf der Tagung behandelten Thematik gliedert sich der Konferenzbericht wie folgt: Grundlegende physikalische Mechanismen, Instrumentierung, Dosimetrie – allgemein, Thermolumineszenz-Altersbestimmung und andere Dosimetrietechniken, Thermisch stimulierte Exoelektronenemission und Spurätz-Dosimetrie, Personen-, Umgebungs- und klinische Dosimetrie, Material-Charakteristiken.

Die auf der Konferenz vorgestellten Arbeiten veranschaulichen den aktuellen Stand auf dem Gebiet der Festkörperdosimetrie (allerdings waren die sozialistischen Staaten unterrepräsentiert). Der Konferenzbericht ist deshalb allen Fachkollegen zu empfehlen, die auf dem Gebiet der Dosimetrie ionisierender Strahlung arbeiten; aber auch für den Festkörper-Physiker sind einige interessante Beiträge enthalten.

K. Hübner

Kristallphysik

S. Haussühl. Leipzig: VEB Deutscher Verlag für Grundstoffindustrie 1983. 433 S., 70 Abb., 27 Tab., umfangr. Verz. deutschsprach. und engl. Lit. (107 z. T. mehrbänd. Bücher und Tabellenwerke, 55 Zeitschriftenartikel); Broschur M 58,–

Mit diesem Buch will der Autor die Besonderheiten des kristallinen Zustandes der Stoffe vermitteln, die phänomenologischen Aspekte unter dem Einfluß der Symmetrie nahebringen und gleichzeitig – wie im Vorwort zum Ausdruck kommt – die praktischen Gesichtspunkte zur Beobachtung und Messung der Kristalleigenschaften beleuchten. Demnach wird das Konzept einer Kristallphysik im klassischen Sinne verfolgt, die sich von dem heute als „Festkörperphysik“ bezeichneten Gebiet vor allem dadurch unterscheidet, daß auf eine realstrukturelle und quantenphysikalische Interpretation der Phänomene weitgehend verzichtet wird. Deshalb werden beispielsweise die Grundlagen der Festkörperelektronik einschließlich der Halb- und Supraleitung oder das Prinzip der Festkörperlaser und -detektoren nicht erörtert. Diesbezügliche Informationen findet man im zitierten Schrifttum.

Für das Verständnis der Darstellung sind neben der Beherrschung kristallographischer Grundlagen solide Kenntnisse in Physik notwendig. Gruppentheoretische Grundlagen sowie die Grundzüge der Vektor- und Tensorrechnung sollten bekannt sein, obwohl einleitend bzw. im Anhang des Buches die für die Kristallphysik wesentlichen mathematischen Grundlagen dargelegt werden.

Der kristallphysikalische Inhalt des Buches ist grob in „tensorielle“ und „nichttensorielle“ Kristalleigenschaften eingeteilt. Die tensoriellen Phänomene sind nach mathematischen Gesichtspunkten weiter untergliedert, und zwar sind jeweils diejenigen zusammengefaßt, die sich mit einem Tensor gleicher Stufe beschreiben lassen. So werden die skalaren Größen Dichte und spezifische Wärme (-kapazität) als „Tensoren 0.-Stufe“, die pyroelektrischen und verwandten Effekte als „Tensoren 1.-Stufe“ und dielektrische, ferroelektrische, magnetische und optische Eigenschaften sowie die elektrische und Wärmeleitfähigkeit unter „Tensoren 2.-Stufe“ behandelt. Piezoelektrischer und elektrooptischer Effekt, der Faraday- und Hall-Effekt sowie nichtlineare Erscheinungen, dielektrische Suszeptibilität und elektrische Transportphänomene werden als „Tensor 3.-Stufe“ behandelt. Der Abschnitt „Tensoren 4.-Stufe“ umfaßt im wesentlichen die Elastomechanik. Im Abschnitt „Tensoren höherer Stufe“ werden weitere aktuelle festkörperphysikalische Effekte erörtert.

An „nichttensoriellen“ Eigenschaften werden Härte und Festigkeit und damit zusammenhängende Kristalleigenschaften behandelt.

Eine Fülle von interessanten Beispielen sowie bewährten und modernen Untersuchungsverfahren lockert die in den Grundlagen abstrakte Darstellung auf.

Es ist zu empfehlen, künftige Auflagen vor allem hinsichtlich der elektrodynamischen Grundlagen konsequent auf die international vereinbarten Begriffe, Symbole, Maß- und Einheitensysteme umzustellen. Auf diese Weise werden unübliche Darstellungen – wie beispielsweise bei den *Maxwell*schen Gleichungen – und fehlerhafte Formulierungen – wie bei der *Lorentz*kraft – sicher vermieden.

Dennoch darf das Buch – gemessen am Konzept der klassischen Kristallphysik – als aktuelle Darstellung gelten, die für Kristallographen, Geowissenschaftler, Physiker und Chemiker sowie werkstoffwissenschaftlich tätige Ingenieure und Studenten höherer Semester entsprechender Fachrichtungen sehr nützlich ist.

H. A. Schneider

Fragen der mathematischen Zuverlässigkeitstheorie

Moskau: Radio i Svjaz 1983; 376 S., 30 Abb., 2 Tab. (russisch) Rbl. 3,–; M 16,25

Das vorliegende Buch ist dem noch jungen Spezialgebiet der Zuverlässigkeitstheorie gewidmet. Durch 6 international anerkannte sowjetische Spezialisten werden der aktuelle Stand und die Richtungen der Zuverlässigkeitstheorie dargelegt. Die Anwendung mathematischer Methoden bei der Untersuchung der Zuverlässigkeit dient einerseits der Präzisierung der Begriffe und Probleme und andererseits der quantitativen Erfassung der Zuverlässigkeit von Produkten sowie der Modellierung von Bearbeitungsprozessen unter der Zielstellung der Erhöhung ihrer Zuverlässigkeit.

Der Teil 1 (3 Abschnitte) wurde von *A. D. Solov'ev* verfaßt und stellt eine elegante Einführung dar. Hier ist auf vergleichsweise geringem Umfang viel Stoff untergebracht, der die für die Zuverlässigkeitstheorie wichtigsten Wahrscheinlichkeitsverteilungen und Grenzwertsätze umfaßt und auch *Markov*sche Prozesse einschließt. Aufbauend auf diesen mathematischen Modellen werden Fragen der Erneuerung von Elementen und der Zuverlässigkeit von Systemen in Abhängigkeit von ihrer Struktur behandelt. Ausgehend von Grenzwertsätzen werden zweiseitige Ungleichungen für Kenngrößen angegeben, die für die Praxis von großer Bedeutung sind. Der 2. Teil (6 Abschnitte) wurde von *Ju. K. Beljaev* und *I. I. Kovalenko* geschrieben und enthält die Anwendung statistischer Methoden. Hier werden Fragen der statistischen Auswertung von Experimenten, der Prüfung von Hypothesen, der Bestimmung von Konfidenzintervallen und der Schätztheorie behandelt. Ein Abschnitt ist der Modellierung komplexer Systeme durch mehrdimensionale *Markov*sche Prozesse gewidmet. Im 3. Teil (7 Abschnitte) werden von *V. A. Kaštanov*, *E. Ju. Barzilovic* und *I. A. Ušakov* Fragen der Optimierung behandelt. Im Mittelpunkt stehen dabei mathematische Methoden der Bedienung von Systemen ohne Berücksichtigung der Struktur, bei unvollständiger Information über die Zuverlässigkeit der Elemente und beim Auftreten von Kontrollfehlern. Ausführlich wird auch auf die Modellierung von Bedienungsproblemen durch semi-*Markov*sche Prozesse und auf Fragen der optimalen Reservierung eingegangen.

Neben der Ableitung mathematischer Modelle aus den Problemen der Zuverlässigkeit und der Behandlung dieser Modelle wird oft auf Fragen hingewiesen, die über die mathematischen Aspekte hinausgehen (z. B. Entwurf wartungsfreundlicher Produkte, Vervollkommnung technologischer Operationen, Untersuchung der physikalischen Ursachen für einen Ausfall). Das Buch wendet sich an wissenschaftliche Mitarbeiter, die sich mit der Anwendung mathematischer Methoden auf Probleme der Zuverlässigkeit von Produkten, Elementen und Systemen beschäftigen.

G. Kummer

Kristallisation aus Schmelzen

Hrsg.: K. Hein, E. Buhrig. Leipzig: VEB Deutscher Verlag für Grundstoffindustrie 1983. 344 S., 215 Abb., 88 Tab., 403 Lit.; Leinen M 50,–

Im vorliegenden Buch werden in konzentrierter Form Probleme der Makroverteilung von Verunreinigungen und Legierungskomponenten in Eisenmetallen, NE-Metallen, Edelmetallen, hochschmelzenden Metallen, Halbleitern und radioaktiven Metallen bei der Erstarrung aus Schmelzen behandelt. Folgende stoffliche Untersetzung wurde gewählt:

- Definition des Verteilungskoeffizienten
- Gleichgewichtsverteilungskoeffizient k_0
- Verteilungskoeffizient unter Nichtgleichgewichtsbedingungen
- Normalerstarrung und Zonenschmelzen als Verfahren zur Reinigung und Kristallzüchtung
- Experimentelle Bestimmung von Verteilungskoeffizienten
- Tabellenanhang

Ausgehend von der Definition des Gleichgewichtsverteilungskoeffizienten und dessen Berechnung aus thermodynamischen Größen wird neben der Behandlung der Temperatur- und Konzentrationsabhängigkeit umfangreich (auch in bezug auf die Zahl der Graphen) die periodische Abhängigkeit der k_0-Werte von der Ordnungszahl dargestellt. In logischer Verknüpfung schließen sich Betrachtungen über Nichtgleichgewichtsbedingungen und effektive Verteilungskoeffizienten an, wobei der Einfluß von Transportmechanismen besonders herausgestellt wird. Der außerordentlich umfangreichen Datensammlung (136 S.!) von stoffspezifischen Daten, Konzentrationsfunktionen und Nomogrammen im Buchanhang steht ein leider sehr spartanisch gehaltener Beitrag zu den Möglichkeiten und den Fehlerquellen der experimentellen Bestimmung von Verteilungskoeffizienten gegenüber.

Erwartungsgemäß werden in einem gewichtigen Kapitel die Verfahren zur Reinigung mittels Normalerstarrung und Zonenschmelzen detailliert erläutert.

Das Buch ist eine gute Mischung aus Lehrbuch und Nachschlagewerk. Von der zitierten Literatur ist ein reichliches Drittel nach 1970 erschienen, und nur 127 Zitate stammen aus der Zeit vor 1960. Es wendet sich vornehmlich an Studenten und an die im Fachgebiet tätigen Wissenschaftler und Ingenieure. Der kurze Exkurs in die radioaktiven Elemente (1 S.) ist jedoch nicht hinreichend für einen Leserkreis, der sich bevorzugt mit der Anwendung radioaktiver Isotope auseinandersetzt.

J. Flachowsky

Nach dem Atomschlag

Ein Sonderbericht von Ambio, veröffentlicht unter der Schirmherrschaft der Königlich-Schwedischen Akademie der Wissenschaften, hrsg. von J. Peterson, dt. Ausg. hrsg. von K. Lohs. Frankfurt/M.: Pergamon Press 1983. 271 S., zahlr. Abb., Tab. u. Lit., mit einem Geleitwort von H. Klare, einem Vorwort zur dt. Ausgabe von K. Lohs, einem Glossarium und biographischen Angaben zu den Verfassern; geb. DM 19,80

Bei diesem außergewöhnlichen Buch handelt es sich um eine wissenschaftliche Studie über die sozialen und ökologischen Folgen eines Nuklearkrieges. Kompetente Fachvertreter aus der UdSSR, den USA, Ost- und Westeuropa (darunter aus der DDR *Klaus Wetzel*) sind die Autoren von 15 übersichtlich gegliederten Einzelbeiträgen, die meist einheitlich mit einem knappen instruktiven Einleitungstext, einer Zusammenfassung, Literatur und Anmerkungen versehen sind.

Im ersten Kapitel werden die gegenwärtigen Kernwaffenarsenale, Waffentypen und taktischen Kernwaffen und deren mögliche Weiterentwicklung bis in die 90er Jahre beschrieben. Der zweite Beitrag „Die Wirkung von nuklearen Waffen" enthält neben vielem anderen Tatsachenmaterial zu den Katastrophen von Hiroshima und Nagasaki. Die Ambio-Beratergruppe entwickelt dann im 3. Kapitel ein ganz konkretes Szenarium zum Thema „Wie ein Nuklearkrieg ausgetragen werden könnte" für den Beginn „an einem Wochentag Anfang Juni 1985 11 Uhr", wobei ein Verbrauch an Megatonnen-Sprengkraft von weniger als der Hälfte der 1985 vorhandenen angenommen wird. Darauf nehmen die folgenden Kapitel Bezug, sie lauten: 4. Epidemiologie: Die Zukunft heißt Krankheit und Tod, 5. Medizinische Auswirkungen der Strahlung nach einem atomaren Weltkrieg, 6. Die Atmosphäre nach dem Atomschlag – Mittagsdämmerung, 7. Auswirkungen auf die weltweite Süßwasserversorgung, 8. Auswirkungen auf die Ökosysteme der Ozeane, 9. Auswirkungen auf die Landwirtschaft, 10. Auswirkungen der ionisierenden Strahlung auf das Leben in der Natur, 11. Ökonomische Konsequenzen: Zurück in die graue Vorzeit, 12. Globale Auswirkungen auf die Nahrungsmittelversorgung, 13. Auswirkungen auf das Verhalten des Menschen, 14. Schlußfolgerungen, 15. Konsequenzen eines „begrenzten" Nuklearkrieges in Ost- und Westdeutschland.

Die so zahlreichen erschütternden Fakten und Resultate wurden von den Autoren in der Überzeugung vermittelt, „daß eine realistische Einschätzung der möglichen Konsequenzen eines mit Kernwaffen geführten Krieges für Mensch und Umwelt dazu beitragen könnte, eine derartige Katastrophe abzuwenden". Die Quintessenz lautet: „Die Auswirkungen des ... Kernwaffenkrieges wären allgemein und furchtbar. Dennoch würde es wahrscheinlich Überlebende geben. Ihr Schicksal wäre jedoch extrem ungewiß" und durch die tiefgreifend langfristig veränderte Umwelt so „beeinträchtigt, daß der Kampf der Überlebenden um eine Erholung von der Katastrophe beinahe aussichtslos sein könnte".

Hier erfährt der Leser ganz konkret, „was sich anbahnt". Er wird nicht nur zum Nachdenken angeregt, sondern unentrinnbar zur Reflexion gezwungen – und findet keine Ruhe mehr.

K.-P. Dostal

Reliability of Reactor Pressure Components Proceedings of a Symposium, Stuttgart 21–25 March 1983

Vienna: IAEA 1983. STI/PUB/645. 415 S., div. Abb. u. Tab., kart. ÖS 820,–

The symposium was organized by the IAEA in co-operation with the "Materialpruefungsanstalt (MPA) der Universität" (Stuttgart, F.R.G.). It was attended by 158 participants and 53 observers from 23 countries and five international organizations. The proceedings include 18 (invited) papers and 25 extended abstracts. The aim is to inform about recent developments in building and operating nuclear power stations of proven types.

Primary Circuit: *H. Schenk* et al. (Kernkraftwerk Obrigheim, F.R.G.) reports on components after 100,000 hours of operation. A high availability (84%) has been achieved by careful maintenance, by automated ultrasonic testing of the reactor vessel, by TV ccanning of the cladding, by an extensive eddy-current testing of the steam generator tubes, by reducing the fast neutron flux at the vessel wall, and by inhibiting corrosive processes. *S. H. Bush* (Battelle Pacific NW Lab., Richland, U.S.A.) deals with failures of piping. *R. Saglio* et al. (CEA Saclay, France) discusses (in French) detection and characterization of flaws by in-serve inspection (ISI).

Steam Generators made of INCONEL 690 (*C. Buchalet*, Framatome, Paris, France) or INCOLOY 800 (*G. Schücktanz* et al., Kraftwerk Union, Erlangen, F.R.G.) mainly suffer from corrosion of the tubing. Anti-corrosive measures (water chemistry, de-aeration) and the latest improvements in ultrasonic (US) and eddy-current (EC) testing are described.

Fabrication and Materials: *S. Onodera* (Japan Steel Works, Tokyo) reports on the quality improvement of heavy steels and their welds by fewer weld deams (better inspectability), by change of cast steels to forged steels, by use of narrow gap welding processes, and by change of welding wire coating from Cu to Al or Cr. A new steel-making process combines the deoxidation of the steel by gaseous CO reaction with the subsequent addition of Al for grain refinement and leads to better weldability and toughness (*F. Hochstein* et al., Klöckner, Osnabrueck, F.R.G.). *M. Brumovsky* (Skoda, Plzen, ČSSR) deals with the influence of Cu- and P-"catalyzed" neutron embrittlement on the hypothetical lifetime of reactor vessels.

Fracture Mechanics of piping is studied experimentally (*K. Kussmaul* et al., MPA, Stuttgart, F.R.G.) and interpretatively (*F. Hochstein* et al., KWU, Erlangen, F.R.G.).

Flaw Detection and Sizing in ISI by modern US methods such as amplitude (phase) – locus – correlation, use of electronically steered and focussed US-probes as well as holography and synthetic aperture focussing technique, respectively, is carried out by *P. Höller* et al. (Fraunhofer-institut, Saarbruecken, F.R.G.). According to *S. Crutzen* (CEC, Joint Research Centre, Ispra) much effort has to be devoted to calibrate non-destructive testing (NDT) with the help of simulating test specimens equipped with artificial defects.

Pumps and Valves and Pipe Cracking as well as **ISI Data Acquisition, Storage and Retrieval with Special Emphasis on the Use in Data Banks for Component Reliability** complete the scope of the symposium.

As a broad spectrum of scientific and engineering activities are required, a reviewing article encompassing the necessary interrelations between them would have raised the value of the whole. Nevertheless, reading the booklet ensures far more than a mere approach for metallurgists, designers, safety engineers and interested material researchers, electronical engineers, physicists and water chemists.

W. Görner

Uranium Extraction Technology (Current practice and new developments in ore processing)

Paris OECD: 1983. 270 S., ISBN 92-64-12397-0; kart.

Die Technologie der Verarbeitung von Uraniumerzen unterscheidet sich stark von der Aufbereitung und Verhüttung anderer Erze, da Methoden der klassischen (physikalischen) Aufbereitung oder Voranreicherung wenig angewandt werden. In den meisten Fällen wird das Erz hydrometallurgisch verarbeitet, um Konzentrate herzustellen. Diese Ausnahmestellung des Uraniums (die nur von Gold und Vanadium geteilt wird) führte zur Entwicklung von technologischen Verfahren, die in der jetzigen Zeit ihrerseits für die Gewinnung anderer Metalle interessant zu werden beginnen. Die vorliegende Veröffentlichung gibt einen systematisch geordneten Überblick des internationalen Standes der Technik, der von einer Gruppe von Spezialisten erarbeitet wurde. Dabei werden sowohl die in der Praxis üblichen und bewährten Verfahren als auch neuere Entwicklungen vorgestellt. Die Veröffentlichung wendet sich nicht nur an Spezialisten, sondern ist für einen größeren Kreis von Technikern und Wissenschaftlern geschrieben.

Der erste Abschnitt behandelt die Geologie der Erkundung der Uranium-Lagerstätten, geordnet nach genetischen Typen (20 S.). Der 2. Abschnitt (92 S.) stellt die Technologie der Verarbeitung dar; Unterabschnitte: Erzvorbereitung (Zerkleinern, Anreichern), Laugung, Konzentration und Reinigung, Konstruktion von Rückstandsbecken, Uraniumgewinnung aus Meereswasser. Bemerkenswert sind die Angaben über die Trennung von Feststoff und Flüssigkeit in Hochleistungs-Eindickern, über neue Ionenaustauscher, über die Beseitigung von Radium und anderen Metallen. Der 3. Abschnitt behandelt kurz (18 S.) Fragen der ökonomischen Bewertung und ihres Einflusses auf die Wahl des Verfahrensganges. Zu jedem Abschnitt wird die neuere Literatur angegeben.

Es folgen ausführliche Berichte über den Stand der Uraniumindustrie in den einzelnen Ländern (125 S.). Die Lagerstätten, hauptsächliche Eigenschaften der Erze, Verfahrensgang, technische und ökonomische Kennziffern werden in Einzelheiten mitgeteilt. Am Ende der einzelnen Länderberichte folgen Literaturhinweise. Folgende Länder haben Berichte gegeben: Argentinien, Australien, Brasilien, Canada, Dänemark, Frankreich, Gabun, Bundesrepublik Deutschland, Namibia, Niger, Portugal, Südafrika, Spanien, Schweden, USA.

Der Band wird mit der Liste der Teilnehmer der Arbeitsgruppe abgeschlossen.

Die Veröffentlichung wird von den Fachleuten des Bergbaus und der Aufbereitung als gute Zusammenfassung des internationalen Standes mit Dank aufgenommen werden.

K. H. Köhler

Safety and Efficacy of Radiopharmaceuticals

Hrsg.: K. Kristensen u. E. Nørbygaard. Boston — The Hague — Dordrecht — Lancaster: Martinus Nijhoff Publishers 1984. 383 S., div. Abb., Tab., Lit.; geb. Dfl. 145.00; US$ 54.00

Die als Band 4 der Serie „Developments in Nuclear Medicine" (Hrsg.: *P. H. Cox*) erschienene Publikation faßt vom Ersten Europäischen Symposium über Radiopharmazie und Radiopharmaka (Elsinore, Dänemark, 27.—30. März 1983) die Übersichtsreferate sowie Einführungsbeiträge einer Paneldiskussion zusammen. Dabei werden Radiopharmakaentwicklung, -herstellung und -kontrolle systematisch und umfassend von den verschiedenen Standpunkten aus resümiert. Die drei Teilgebiete werden ergänzt durch Vorwort, Autorenverzeichnis und Sachregister.

Teil 1 mit insgesamt 22 Beiträgen und aktuellen Literaturangaben behandelt Radiotracerdesign, vorklinische Prüfung und klinische Testung von Radiopharmaka einschließlich organisatorischer, arzneimittelrechtlicher und ethischer Fragen. Radiopharmaka-Arzneimittel-Wechselwirkungen werden ebenfalls besprochen.

Die Besonderheit der in der Klinik stattfindenden Präparation von radioaktiven Arzneimitteln und darauf gerichtete Empfehlungen bzw. Vorschriften für die Laborausstattung zur adäquaten aseptischen Arbeitsweise bei entsprechendem Strahlenschutz sind Gegenstand der drei Beiträge des zweiten Teils der Monographie.

Im Teil 3 wird die Qualitätskontrolle von Radiopharmaka, die mit Hilfe von Generatoren und Kits präpariert werden, vom Standpunkt des Produzenten, der klinischen Einrichtung und der für die Überwachung von Arzneimitteln zuständigen Institution diskutiert.

Die Monographie ist für alle, die sich mit der Entwicklung, Prüfung und Herstellung von Radiopharmaka befassen, sicherlich eine wertvolle Informationsquelle.

B. Johannsen

Isotope und ihre Anwendung in den Geowissenschaften, in der Bergbausicherheit und im Umweltschutz

Freiberger Forschungshefte C 388 (Teil I) und C 389 (Teil II). VEB Deutscher Verlag für Grundstoffindustrie Leipzig: 1983/84 Teil I: 231 S., 98 Abb., 19 Tab., 199 Lit.; geheftet M 70,— ; Teil II: 297 S., 65 Abb., 61 Tab., 385 Lit.; geheftet M 90.—

Die vorliegenden beiden Hefte beinhalten die Vorträge des internationalen Isotopenkolloquiums zum Berg- und Hüttenmännischen Tag 1982 in Freiberg. Im Teil I werden in Form von 18 Einzelarbeiten insbesondere die Möglichkeiten herausgearbeitet, die sich durch den Einsatz isotopenhydrogeologischer Methoden und deren Interpretation ergeben. Dabei zeigt sich, daß die Umweltnuklide D, T, ^{13}C, ^{14}C, ^{18}O, ^{32}Si u. a. grundsätzlich geeignet sind, sowohl in den frühen Erkundungsphasen wertvolle Hinweise zum hydrogeologischen Modell zu liefern als auch spezielle Probleme eines mit klassischen Methoden gut untersuchten Gebietes zu lösen. Auch die Vorteile einer kombinierten Anwendung von Umweltnukliden unter realen geologischen Bedingungen werden aufgezeigt. Sie sind z. T. sogar notwendig, um zu den gewünschten Aussagen zu gelangen.

Der Teil II (24 Einzelarbeiten) umfaßt im wesentlichen die Untersuchungen des festen Bereiches. Einen Schwerpunkt bilden dabei die K/Ar-, Rb/Sr- und Pb- sowie ^{34}S-, ^{18}O- und ^{13}C-Isotopenuntersuchungen im magmatischen Bereich, wobei allgemeine Überlegungen und auch konkrete Einzelergebnisse diskutiert werden. Außerdem werden Arbeiten theoretisch-methodischer Art und solche über das sedimentäre Geschehen, über Lagerstätten und über Umweltprobleme vorgestellt.

Insgesamt geben die beiden Hefte für den Fachmann sowohl einen guten Überblick über den Entwicklungsstand auf diesem Spezialgebiet als auch über die Möglichkeiten und Grenzen der einzelnen Institutionen. Für nicht auf diesem Gebiet Tätige können die übersichtlich und informativ dargelegten Ergebnisse eine Anregung sein, über die Anwendung von Isotopenmethoden auf ihrem Spezialgebiet nachzudenken.

P. Harting

Information Technology and the Computer Network Proc. of the NATO Advanced Study Institute on Information Technology and the Computer Network, Bonas, France, 21. Aug.—2. Sept. 1983

Hrsg.: K. G. Beauchamp. Berlin — Heudelberg: Springer-Verlag 1984. 271 S., 51 Abb., 4 Tab., 145 Lit.; karton. DM 92,—; US $ 34,40

Der Tagungsband enthält eine Zusammenstellung von 27 Beiträgen zum Themenbereich rechnergestützter Informationstechnologien und Datenkommunikationsnetzen – insbesondere Breitbandkommunikation, verteilte Verarbeitung, Informationsübertragung, Sicherheit und Datenschutz, Netzmanagement, dienstintegrierte digitale Netze, theoretisch-technologische Entwicklungen sowie Nutzungs- und Leistungsanforderungen an die Datenkommunikation. Anhand ausgewählter Projekte bemühen sich die Autoren, den gegenwärtigen Stand rechnergestützter Informationstechnologien aufzuzeigen, den gestiegenen Bedarf nach neuen Informationstechnologien und Informationsleistungen darzustellen und auf künftige Problemstellungen und Richtungen zu verweisen.

Der Schwerpunkt der Beiträge ist den Arbeiten gewidmet, national und international kompatible Datenkommunikationsnetze zu schaffen, die eine umfassende Nutzung verschiedener Leistungen und Dienste ermöglichen. Die Problematik einheitlicher Schnittstellen, Protokolle und Standards, insbesondere auf der Basis des ISO-Referenz-Modells für öffentliche Datenkommunikationsnetze (Open systems interconnection/*W. A. McCrum* und Issues in multimedia computerbased message system design and standardization/*J. J. Garcia Luna Aceves; F. F. Kuo*), wird deshalb ausführlich behandelt.

Des weiteren werden Projekte zur Kopplung lokaler Netze – z.B. über Satellit im Rahmen des UNIVERSE-Projekts in Großbritannien (The Universe Project/*C. J. Adams*) – sowie die Kopplung lokaler Netze an öffentliche Datenkommunikationsnetze beschrieben. Der gegenwärtig erreichte Stand dokumentiert sich im Leistungs- und Nutzungsumfang beschriebener Projekte. Hier sind das internationale Rechnerverbundnetz SITA auf dem Gebiet des Lufttransportwesens (SITA advanced telecommunications services/*G. Giraudbit*) und das Netz SERC zum Verbund von Rechnern im F/E-Bereich in Großbritannien (The SERC Network – Its history and development/*P. Bryant*) hervorzuheben. Weitere Beiträge beschäftigen sich mit zivilen und militärischen Sicherheitsfragen bei der Datenkommunikation und dem Datenkommunikationsmanagement.

Der vorliegende Band gibt sowohl für den Spezialisten als auch den Nichtfachmann einen bewerteten Überblick über den Themenbereich rechnergestützter Informationstechnologien und Datenkommunikationsnetze und verweist auf wichtige Literatur und Standards. *H. Lamprecht*

Mikroelektronik
Computer, Roboter und neue Medien erobern die Welt

S. M. Gergely. München/Zürich: R. Piper 1983. 315 S., 82 Abb.; Leinen DM 29,80

Die Mikroelektronik und die Anwendung ihrer Produkte in der Volkswirtschaft bestimmen heute zunehmend das Tempo des wissenschaftlich-technischen Fortschritts. Die Verschmelzung der EDV-, Informations- und der Kommunikationstechnik, das Entstehen neuer Medien und nicht zuletzt die Herausbildung einer Informationsindustrie beeinflussen entscheidend die Produktivität der Herstellungs- und Verwaltungsprozesse und verändern immer mehr das Leben der menschlichen Gesellschaft. Die Vielfalt und die Intensität der technischen und technologischen Veränderungen, die sich mit der revolutionären Entwicklung der Mikroelektronik vollziehen, machen es für jeden Nichtspezialisten immer komplizierter, diese Entwicklung zu verfolgen und ihre möglichen Auswirkungen auf die konkrete Wissensdisziplin einzuschätzen. Der besondere Wert eines Buches der vorliegenden Art liegt gerade in der wissenschaftlich streng fundierten und dennoch allgemein verständlichen Darlegung der Technik und Funktionsweise von Geräten, Anwendungsmöglichkeiten und gesellschaftlichen Auswirkungen der Mikroelektronik.

Stefan M. Gergely hat in seinem Buch die derzeitige Diskussion über die Möglichkeiten, Grenzen und Gefahren der Einführung neuer Medien zusammengefaßt und versucht, einige Hintergründe der mikroelektronischen Revolution unserer Zeit aufzuzeichnen.

Nachdem er einige technische Grundlagen der Computertechnik, der Halbleitertechnik und die Informations- und Telekommunikationstechnik dargestellt hat, widmet er den zweiten Teil seines Buches den Fragen der Anwendung der Mikroelektronik. In diesem Teil werden folgende Bereiche behandelt: Datenbanken, Entstehung und Bedeutung neuer Medien (Bildschirmtext, Teletext, Kabelfernsehen), Satelliten, Mikroelektronik in Haushalt, Freizeit und Unterricht; Mikroelektronik in Verwaltung, Verkehr, Wissenschaft und Kunst; Automatisierung der Produktionsprozesse; Computergrafik.

Im dritten Teil des Buches – Mikroelektronik und Gesellschaft – versucht der Autor, eine Reihe von Fragen zu beantworten, welche für den sozialen Fortschritt von entscheidender Bedeutung sind. Er untersucht die Wechselbeziehungen zwischen der Anwendung der Mikroelektronik und der Arbeitslosigkeit, die Datenschutzprobleme, die Möglichkeiten der Mikroelektronik bei der Lösung der Bildungskrise in den Entwicklungsländern und die Probleme der künstlichen Intelligenz und leitet eine Reihe von Schlußfolgerungen über die Rolle der Mikroelektronik bei den tiefen strukturellen Veränderungen des gesellschaftlichen Lebens in der Zukunft ab.

Seinen Text belebt der Autor durch zahlreiche sehr gut ausgeführte Illustrationen und einige Schemadarstellungen. Neben den 67 Literaturquellen, auf welche er sich unmittelbar im Text bezieht, verweist er am Ende des Buches auf mehrere hundert aktuelle Zusatzquellen aus dem Gebiet der Computer- und Informationstechnik, Telekommunikation und Anwendung der Mikroelektronik. Ein exakt erarbeitetes alphabetisches Sachregister erleichtert dem Leser einen schnellen Einblick in den ihn interessierenden Bereich.

Das Buch ist geeignet für einen breiten Leserkreis als wissenschaftlich präzise und leicht verständliche Einführung in die Probleme der Mikroelektronik. Es bleibt zu hoffen, daß analoge Sachbücher zu ähnlichen wissenschaftlichen und technischen Entwicklungen unserer Zeit ebenfalls einen großen Leserkreis gewinnen. *L. Jankowski*

Galileo Galilei – Prozeß ohne Ende
Eine Biographie

A. Fölsing. München/Zürich: R. Piper 1983. 500 S., 49 Abb.; Leinen DM 39,80

Auf den ersten Seiten eines Lehrbuches der Physik lernt heute jeder Schüler und Student Begriffe und Gesetze kennen, die ihm fast selbstverständlich vorkommen: Geschwindigkeit und Beschleunigung, das Trägheitsprinzip, die Fallgesetze und die Wurfparabel. Sie alle können auf einen Großen der Physik, ja der Naturwissenschaft überhaupt, auf *Galileo Galilei* zurückgeführt werden, der den experimentellen Dialog mit der Natur in die Naturwissenschaft eingeführt hat, die wohlüberlegte Fragestellung, um Unbekanntes zu ergründen und – wenn möglich – in der Sprache der Mathematik auszudrücken.

Der Autor *Albrecht Fölsing*, sachkundiger und an philosophischen/historischen Problemen interessierter Wissenschaftsjournalist, hat aufgrund umfangreicher Studien und Materialsammlungen ein Werk über das Leben des großen *Galileo Galilei* zusammengetragen, das „ohne Legenden" den Menschen zu erfassen versucht und mit manchen falschen Auffassungen aufräumt. Der eingangs erwähnte Schüler oder Student lernt auch, daß *Galileo Galilei* seine Fallversuche auf dem schiefen Turm zu Pisa ausgeführt oder daß er, mit dem Fuße aufstampfend, vor dem Inquisitionstribunal erklärt hätte „. . . und sie dreht sich doch!" Dies und manches andere sind Legenden, viele Jahrzehnte nach den tatsächlichen Ereignissen erfunden und seither weiterverbreitet. Daß *Galileo Galilei* die Fallgesetze durch ein reines Gedankenexperiment entdeckte und daß er einer der Begründer der italienischen Accademia dei Lincei (1624) – Akademie der Luchse – war, die noch vor der berühmten Royal Society in London (1662) gegründet wurde, sollte auch bekannt sein oder bekannt gemacht werden.

Es setzt uns heute in Erstaunen, wieso Aristoteles 2000 Jahre lang eine so große Autorität in allen Fragen der Natur und der Philosophie sein konnte, da doch keine einzige Schrift von ihm selbst herausgegeben wurde und seine Überlieferer manchen Widerspruch weitergereicht haben. Vielleicht hätte man an einer geeigneten Stelle einige Sätze über die *Aristoteles*sche Betrachtung der Bewegungen, wie sie *Galileo Galilei* vorfand, verwenden sollen: die 4 Arten der Bewegung nach *Aristoteles* oder die Impetuslehre der Scholastiker. Auch die Philosophie *Spinozas* sollte Erwähnung finden, die in einem besonderen Maße versuchte, die innere Harmonie von Ursache und Wirkung zu verstehen, und die der cartesischen Physik mit zum Durchbruch verhalf. Es ist heute kaum vorstellbar, in welcher geistigen Umwelt *Galileo Galilei* sich damals befand: Der erste Philosoph an der Universität Pisa, Professor *Giulio Libri*, hat u. a. seinem Großherzog *Cosimo* auseinandergesetzt, daß es die ihm gewidmeten Sterne schon aus Gründen der Logik gar nicht geben könne, und dieser Professor starb, ohne jemals durch ein Fernrohr geblickt zu haben. Was würde dieser erste Philosoph heute von den Bemühungen der Physiker auf dem Gebiet der Elementarteilchen halten? Auch *Bertolt Brecht* hat diesem Philosophen ein Denkmal in seiner Weise gesetzt: „Herr *Galilei*, bevor wir Ihr berühmtes Rohr applizieren, möchten wir um das Vergnügen eines Disputes bitten. Thema: Können solche Planeten existieren?"

Galileo Galilei – Prozeß ohne Ende? Ja, ohne Ende – gerade eben werden Pressemeldungen des Vatikans zufolge Dokumente über den Prozeß in einer Sammlung von 147 Schriftstücken zum erstenmal veröffentlicht, von denen 6 bisher völlig unbekannt sind. Dies wird wieder neue Einblicke möglich machen.

Galileo Galilei lebte nach dem Prozeß vom 22. 6. 1633 noch $8^1/_2$ Jahre, erblindete und starb am 8. 1. 1642. Auf seinem Grab ist zu lesen: „Er verlor sein Augenlicht, weil es für ihn in der Natur nichts mehr zu sehen gab". Er ist der Begründer der modernen Naturwissenschaft und viele, sehr viele sollten sein Leben, sein Wirken und seine Konflikte kennenlernen. Das vorgelegte Buch ist eine ausgezeichnete Hilfe und Grundlage für ein solches Kennenlernen.

H. Hübner

Deutsch-russisches Wörterbuch der Aufbereitung und Verarbeitung mineralischer Rohstoffe

G. M. Bunin, N. K. Kemežuk. Moskva: Russkij Jazyk 1983. 519 S., etwa 25000 Fachbegriffe; geb. Rbl. 4,80; M 24,–

Die zunehmende Verknappung mineralischer Rohstoffe mit hohen Gehalten an nutzbaren Komponenten sowie die immer aufwendiger werdende Gewinnung mineralischer Rohstoffe zwingt die Wirtschaft vieler Länder unter anderem zur verstärkten Nutzung und Aufbereitung armer mineralischer Rohstoffe und zu ihrer rationellen Verwendung. Dadurch haben die Entwicklung neuer und die Vervollkommnung konventioneller Aufbereitungs- und Verarbeitungsverfahren für mineralische Rohstoffe in den letzten Dezennien einen bedeutenden Aufschwung erfahren und zur Herausbildung eigenständiger Industrien und Forschungsdisziplinen geführt. Die damit verbundene Flut wissenschaftlich-technischer Informationen machte die Herausgabe spezieller Wörterbücher für diese Fachgebiete notwendig. So ist das Erscheinen dieses hier rezensierten deutsch-russischen Wörterbuches zu begrüßen. Es enthält zahlreiche Fachbegriffe aus dem Sektor des gesamten Prozesses der Aufbereitung mineralischer Rohstoffe (z.B. Kohle, Torf, Öl- bzw. Brennschiefer, Erze für die Bunt- und Schwarzmetallurgie und Baurohstoffe) von den vorbereitenden bis hin zu den Endstadien sowie des daran anschließenden Prozesses der Verarbeitung der Rohmaterialien. Daneben sind auch Termini zu finden, die die Homogenisierung, Lagerung, Beprobung und Qualitätskontrolle der Ausgangs- und Endprodukte der Aufbereitung, die Vorbehandlung von Konzentraten für die weitere Verarbeitung, die Verkokung und Hydrierung von Kohlen, die Gasreinigung sowie chemische und hydrometallurgische Prozesse abfallarmer Technologien der komplexen Rohstoffverarbeitung betreffen. Ein ca. 5000 russische Fachbegriffe umfassendes Register am Ende des Buches soll das Auffinden häufig gebrauchter oder schwieriger ins Deutsche zu übertragender russischer Termini erleichtern. Das vorliegende Wörterbuch wird nicht nur den im wissenschaftlich-technischen Informationsdienst Tätigen und den Fachübersetzern unentbehrlich, sondern auch Studenten, Wissenschaftlern und Technikern verschiedener Fachdisziplinen des Bergbaus, der Aufbereitung, Metallurgie und Baustoffindustrie zu empfehlen sein. Hin und wieder dürfte sich das Wörterbuch ebenfalls einigen auf dem Gebiet der Isotopen- und Strahlenforschung Arbeitenden als nützlich erweisen, werden doch gegenwärtig unter anderem stabile und radioaktive Nuklide als Tracer bei Forschungen zur optimalen Gestaltung von Aufbereitungs- und Verarbeitungsprozessen von Rohstoffen verstärkt eingesetzt.

Hervorzuheben ist die gründliche Durchsicht; Druckfehler sind bei den deutschen Termini kaum zu finden, und nur selten sind die deutschen Entsprechungen russischer Termini nicht ganz glücklich gewählt (z.B. Vergleichsmäßigung – усреднение). Obwohl sich das Wörterbuch laut Vorwort auch an die in der Baustoffindustrie Beschäftigten wendet, wird der betreffende Nutzer vielleicht einige gebräuchliche Begriffe wie Zement, Gasbeton, Klinker oder Ziegel vermissen, was aber den Wert des Wörterbuches insgesamt nicht mindert.

U. Wand

Lexikon der Kybernetik

Im Auftrag des Zentralinstituts für Kybernetik und Informationsprozesse der Akademie der Wissenschaften der DDR herausgegeben von GÜNTER LAUX

(Übersetzung aus dem Russischen)

A–E

1980. 755 Seiten – 120 Abbildungen – 1 Falttafel – 19 Tabellen – 14,7 × 21,5 cm – Lederin DDR 70,– M; Ausland 70,– DM

Bestell-Nr. 762 1920 (6297/1)

F–L

1981. 642 Seiten – 135 Abbildungen – 15 Tabellen – 14,7 × 21,5 cm – Lederin DDR 60,– M; Ausland 60,– DM

Bestell-Nr. 762 1939 (6297/2)

M–R

1981. 704 Seiten – 123 Abbildungen – 12 Tabellen – 14,7 × 21,5 cm – Lederin DDR 65,– M; Ausland 65,– DM

Bestell-Nr. 762 6019 (6297/3)

S–Z

1981. 589 Seiten – 97 Abbildungen – 16 Falttafeln – 14,7 × 21,5 cm – Lederin DDR 55,– M; Ausland 55,– DM

Bestell-Nr. 762 6027 (6297/4)

Neu:

Ergänzungsband

1985. Ca. 560 Seiten – 97 Abbildungen – 14 Tabellen mit Abkürzungs- und mehrsprachigem Schlagwortverzeichnis – 14,5 × 21,5 cm – Lederin DDR ca. 50,– M; Ausland ca. 50,– DM

Bestell-Nr. 763 4078 (6297/5)

Bitte richten Sie Ihre Bestellungen an eine Buchhandlung

AKADEMIE-VERLAG · BERLIN

DDR-1086 Berlin, Leipziger Str. 3–4

ISOTOPENPRAXIS

Band 21 — 1985 — Heft 5

Contents / Inhalt

►Language of Publication — ►Publikationssprache

Isotopenpraxis is indexed in Current Contents/PC & ES, Science Citation Index, and INIS Atomindex